LIVING MORE SUSTAINABLY

Tools And Tales To Help You Save Our Planet

Elizabeth Kelch

Kelch Publishing

ISBN-13: 979-8667065098
ISBN-10: 1477123456

Cover design by: Art Painter
Library of Congress Control Number: 2018675309
Printed in the United States of America

For my Mom and my Lovey for helping me make this happen.

For mMy kids, and all my loved ones and friends,
for being material for this book and putting up
with my incessant, treehugging ways.

CONTENTS

Preface

This collection of essays is a description of the ways I live sustainably. The stories are of the things I do and have done and they are supported by my explanations of why I do it. I'm pretty opinionated and pushy, so this work began as a how-to and a rant about doing things more sustainably. Each chapter outlines changes that are extreme, significant, or doable beginnings. Its purpose was to show how people can make changes toward more sustainable behavior and life choices and why we all need to.

Sustainability is where I want to focus my writing efforts. As I evolve as a writer, I'm hoping to be less judgmental and more positive in my work. My efforts will, with a bit of effort, advance towards a more gracious and optimistic telling of my stories. When it's time to write a new essay or story, I look around my life and try to find an example of a sustainable living practice I'm already doing. The story of what I'm doing becomes the basis of the essay. My process changed as I found I wanted each of the essays to have more of the personal touch. I wanted this to be how-to, and not so much formal instruction. Instructional doesn't have a very personal touch.

As I wrote more and more of the essays, I found they fell into categories. They seem to match up with the chapters I have been envisioning for the how-to book, with a few minor adjustments, like how I rolled the "My Story" chapter into the "Introduction" chapter. I also found I needed more personal stories as they'd better show how and why I do what I do. I'm not a subject matter expert in sustainability and I don't have the expertise to speak with any authority. But I do know my own story so that's where I focus my writing.

Any description of my journey as a writer has to start with reading Dr. Seuss as a child. I've always read and loved his fun and playful work. I said, "I want to write like him when I grow up". But I wasn't a good reader. I remember thinking, "someone who can't read very well, certainly can't be a writer." Later in life, it was discovered I have a disability slowing my reading tremendously and, boy howdy, was I wrong about being a poor reader. With assistive technology to overcome my disability, I've become a reading machine and, wahoo, I can be a writer. Childhood dreams of writing became available to me.

When I found I was a much stronger reader than I thought, I was ravenous. I read all the stuff I had only dreamed of being able to dive into. Kipling, Hemingway, and Lewis Carroll. Christopher Paolini and Piers Anthony became my guilty pleasures. Emerson, Abby, anybody in Environmental Literature became my mentors. These authors paint pictures with their words, as is true of Dr.

Seuss. He always had and still does for me. I wanted to read classics because I figured they had stood the test of time. Some more recent authors I wanted to emulate are Terry Tempest Williams, Barbara Kingsolver, and Bill McKibben. Their influences incorporated into my work through my use of storytelling. The book evolved from a how-to into the narrative it is. I wanted to have personal stories in it, like these authors have.

I dream of my work being amusing and entertaining. I'd like to give my readers a good laugh and a good think. Sometimes using an acerbic wit makes people laugh and sometimes I use a caricature of myself. Developing my voice has been a process of tempering my own need to grab our society by the shoulders and say "We have to fix this", and trying to be understanding that isn't the best way to motivate change. As my skills and career progress, I want to write so my voice can be the expression of the change we need to save our home.

CHAPTER 1 -

INTRODUCTION:

HINTS OF WHAT'S

TO COME

My Story: How I Got Myself Into This Position

Trying to think back over the years to figure out where my sustainability mindset began, I ponder how I got myself from an ordinary mainstream upbringing in the 70s to the tree hugger I seem to have turned myself into. Sitting at the kitchen table, the place where all meaningful conversations in my family happen, chatting with my mom, she told me she remembered it starting with small stuff at first. My tree hugger personality was beginning to show as long ago as the 80's and 90's. She reminded me of how I wanted solar panels before it was really a feasible means of power and how I rode my bicycle in college because I said it made more sense than a car.

I remember being young and thinking, "I feel just as bad the carrot or rice plant has to die so I can eat, as that the cow or chicken has to die so I can eat." It's cruddy but it's the cycle of life. When I die, something else will live because my body, the organic matter making up my mass, will be added back into the cycle of life. Our planet has worked that way for many millennia and it'll continue for many more millennia to come. Breakdown and decomposition have to happen. If it didn't, we'd be hip deep in... god knows what. That's the problem with landfills.

Back in the 90's, I quit eating meat. I told myself if I didn't eat meat, I still had to eat something, and hopefully that something would be better for me, like more fruits or vegetables, or less "miracles of modern chemistry". After many years of not eating meat and reading environmental science and news of sustainable choices for living and helping our planet remain able to sustain us, I've gone back to eating meat, just a little and only local, sustainably raised. Giving up one food group isn't the answer. Balancing our needs with Mother Earth's ability to provide seems smarter. I also eat local, sustainably raised fruits and vegetables and drink sustainable, fair-trade teas and coffees. I don't want to eat or drink or use anything that causes me guilt for the damage or pain I'm affecting. I'm not always as good at it as I'd like and I am, by no means perfect, but I'm doing what I can.

As I embrace my inner sustainabilitarian, my propensity to extrapolate decisions and ideas all the way out to their logical conclusions, brings me to a difficult place to live in our culture, namely, outside the mainstream. Using the information I have, I infer all the possible consequences or outcomes of a decision and as I make my choices to eat sustainable foods, I feel a need to make other, and eventually all, areas of my life sustainable. I feel as though it's hypocritical of me to eat sustainably but still create excessive waste or continue to encourage plastic production by using it or any number of other unsustainable choices.

I find myself wound up in guilt for not having done more to save the environment. I have to remember what I did today is more helpful than what I was doing yesterday. And tomorrow I will make even better choices that will move us all toward a healthier world. I will do what I can do, one day at a time.

Where Did Sustainabletarian Come From?

"Mom, you're eating meat! I've never seen you do that before". I had been a vegetarian longer than my children had been alive. They were surprised to say the very least. "I won't be finding the nearest McDonald's anytime soon", I told them. Having come through an evolution of food and lifestyle choices, each more sustainable then the last, here is where I am today. My children have labeled me a Sustainabletarian.

Many years ago, I heard someone say "be true to yourself" so I quit eating meat. I had wanted to for a while, but it's hard to go against the grain; to do other than what our culture tells us is "normal". And Americans are meat eaters. When I was growing up, mainstream culture, at least the part of it I was aware of, still labeled the hippies from the 60's "weirdo's", so that's what I learned was the definition of weird. And since those weird hippies were vegetarians, I certainly didn't want to be that when I was a kid. But sometimes it only takes finally listening to something we've heard a million times.

My choice back then was partially a dietary choice. I told myself if I didn't eat meat, I would have to supplement my diet with something and hopefully it would be healthier. And partially it was an ethical choice. I had read some news and investigative journalism on the industrialization of agriculture and had some experience seeing feedlot animals and their treatment. I believe it's wrong to treat any living thing like that. I don't object to animals dying because it's part of the cycle of life. Everything dies. Something has to die so something else can live. I used to say I feel just as guilty about the broccoli dying as the chicken dying. I don't actually feel very guilty about either because I believe it's the way Mother Nature set up our system and I believe it's a good system. What I feel guilty about is the way people treat animals while they're alive. Over time my dietary choice has melded with the ethical choice. It's healthier for me to eat food raised sustainably. I'm still choosing not to eat unhealthy foods and replacing those foods in my diet with healthy ones.

As my children came along, I found I didn't want to feed them unhealthy food either. It seemed they had the same dietary needs as anyone else they just needed the food mushier and smaller. So, whatever we were having for dinner got mushed, by blender at first and a fork later, and fed to the children. I feel better knowing my loved ones are eating healthy food. I wanted them to eat more fruits and veggies too. When it came time for the kids to start eating finger foods, I'd put them in the highchair and, instead of putting boxed cereal on their tray, I used frozen peas and carrots. They became snack food in our house. I reasoned the colors were better, the frozen temperatures probably felt good on their teething gums, and learning to pick them up was as good for developing their motor skills because the peas rolled around the tray.

In more recent years my tree hugger ways have extended well beyond my food choices. I've collected canvas bags over the years as souvenirs from my travels. As our society's thoughts turned more to sustainability, everybody started reusing bags more, so I got out all my canvas souvenirs and started dragging them to the stores with me. Now every time a cashier tries to put something in a plastic bag, I tell them "Save the plastic".

I'm not much of a straw user, so I started saving plastic by handing the straws back to the server right away. I have no objection to straws, my problem is with sigle use items, and unused items going in the trash needlessly. Every time I order a drink I say "no straw please". I know it's a minor contribution, but things add up. How many straws have you been given in your life that ended up unused and, still wrapped, went in the trash? Waitresses and bartenders always lay straws on the table or put them in your drink. If you aren't a straw user, hand it back to them.

A few years ago, I started making all my own personal care

products. I had expected a big ordeal involving lots of expensive ingredients, which is probably why it took me so long to try it. I couldn't have been more wrong. Everything is made from reasonably priced products found at the grocery store using recipes readily available online. Replacing shampoo with a baking soda water mixture and replacing conditioner with apple cider vinegar turned out to be one of the easiest product alternatives I found. It's called the "no-poo" method. I also use 100% aloe gel as my setting lotion/mousse. I have never found a commercial hair product I was really happy with. It was always too sticky, too stiff, didn't control the frizz well enough, etc., but using these hair care products has my hair healthier and more manageable than I can remember it ever being. I wish I had discovered this hair care regime years ago. To say nothing of the money I'm saving not buying all those products that never really did the job very well anyways.

The skin moisturizing lotion I make is a simple recipe made from just a few inexpensive ingredients and even non tree hugging types, say they really like my lotion. When I read the ingredient list on many skincare products, they all have alcohol or petroleum products. I wonder who thought it was a good idea to use the same stuff we fuel our engines with, on our skin, in cosmetics and healing ointments.

Many of us have a cupboard full of plastic food storage containers, in all states of disrepair. Consider the plastic container that's pink and bubbled from microwaving tomato sauce in it. Does it stand to reason, if there's food in your plastic, then there's plastic in your food. Do you want to consume plastic particles? Now consider all the glass jars you've tossed in the recycle bin over the years. Why not store foods in safer, glass containers you already have instead of buying chemical-laden, environmentally destructive, plastic ones. I understand there are times when one doesn't want to have to worry about breakage with a glass container. We have 2 bins for storage containers in our kitchen; one for plastics and one for glass.

My kids have heard a lot about sustainable foods and lifestyles over the years, so when I added sustainably raised meats back into my diet, I tried to explain why I was choosing to eat meat again. I told them about my original choice to not eat meat and how those same reasons make eating sustainably raised meats safe and healthy. I told them about why I choose to use less plastic, make my own personal care products and any number of other choices I make, I believe them to be more sustainable. None of this is information the kids hadn't heard plenty often, so I told them "I guess I'm not a vegetarian anymore, I'm a sustainabletarian".

Sustainability Is About...

An advertisement I saw for the Shell Company said there is 230 years of natural gas available. My first thought was "what do we do in 231 years". Why don't we do that now? Why put money and energy into getting gas out of the ground when we could put that money and energy into research and development that would provide us power indefinitely. Why put money and energy into getting gas out of the ground when the removal of fossil fuel and consumption of oil has proven to be detrimental to the human race's long-term quality of life.

Sustainability is about not doing anything we can't continue to do forever. It means not behaving in a manner which, if continued interminably, would, at a point, become impossible.

In the late 70s or early 80s solar power companies were getting started and they put satellites in space with the best technology of the day. Those satellites are still orbiting with solar technology from 30-40 years ago, and still operating. How much solar power has advanced in those 30-40 years. How much more power is available from today's solar panels, than is provided by those satellite's solar panels manufactured 30-40 years ago. I don't have any specific numbers or technical specifications and quantifying it all isn't really what's important. What is important is that we realize the huge investments in those 30-40 years. News reports tell us solar power's capacity has grown by leaps and bounds and that's the point.

In my tiny off grid camper, I have a solar setup, a battery about the size of a car battery, built into a unit with an inverter on top. The battery and inverter with multiple connections for multiple types of connectors is a unit on wheels and I have two solar panels to charge it, each is 2-3ftx2-3ft. I work from home, so I have a computer, Internet hot spot and phone to run all day for work. I have used my solar setup to run all my electronics for an entire ten-day trip in my tiny off-grid RV.

With all the work being done on solar and wind power generation the work being done on battery storage capacity is truly groundbreaking. We can make all the power we can conceive of with the sun and wind, but if we can't store it, it won't do us any good. Imagine yourself sitting at home on a calm evening... There's no sun or wind so it's a lot darker and quieter than you planned for. Using the energy Mother Nature provides is well and good but only when she is providing the type of energy you have the means to harness. She provides an abundance of energy but it may not be available 24/7. We have to have a means of storing energy for when she's, shall we say, less energetic.

Planning for the future is not just for when you know the refrigerator is going to go kaput in the next few years, or for retirement savings. It's planning to ensure our descendants will have this same warm, life-nourishing, blue green marble to live on. Extrapolate everything you do to the N'th degree. If I throw my trash away like this will my descendant's, generations from now, be able to dispose of the trash the same way. If I take resources out of the earth the way I am now will my descendants, generations from now, be able to do the same. In years past our ancestors may not have had as many choices about how they lived or put as much forethought into considering the future. They may have had limited resources and limited understanding of the consequences of their actions, but today we have plenty of choices and knowledge, enough to consider our descendant's prospects.

A neighbor of mine is very concerned about the next election. He's very distraught, a certain candidate might get into the White House and would then be able to control the Supreme Court, thereby paralyzing the legislative branch. I'm not proposing any of these theories are correct or incorrect. Being worried about the next four years seems very shortsighted. Even to say 200+ years of American policy is at fault, and it might take as much as 200 years to straighten out those policies, seems shortsighted. This is much bigger than any one country or any one election.

Worldwide policies for the past several hundred years have gotten us into the problems we're experiencing now. One candidate and one election can't make enough difference to overcome age old, culturally engrained dogmas? It took people worldwide to create the situation we are now facing, and it will take people, each and every one of us, doing everything in our power, to fix the condition of our home.

People use the argument "this is the way we've always done it". These are the folks who may not yet have seen the T-shirt that says, "those who forget history are doomed to repeat it". Living so unconsciously is not learning the lessons afforded us by our mistakes. People who live so unconsciously are the people who don't evaluate each choice and its consequences. The way we do things now may be the way things have been done in our cultural memory, but that only goes back three, maybe four generations, through the people who are still living. Beyond that, its ancient history, and people often don't think of it as something relevant in our day-to-day lives. Sustainability is about not doing things the way they've always been done blindly and living consciously. Instead, consider everything we do, reevaluate every decision to be made about how we behave, no matter how small. Doing things the way they've always been done has led us to the state we're in.

The philosopher René Descartes proposed to question everything. He questioned everything using radical doubt. He questioned all manner of what the average man might consider facts. Whether or not the strawberry was red, or if indeed it existed, whether or not water is wet, whether or not he had a body. And with each question he found the knowledge of the answer could be explained away as a figment of imagination, trickery or inaccuracy. He questioned all the way back to his existence but this at the very least could be proven because the very action of doubt supplied verification of the existence of one's mind. There must be a thinking entity for there to be a thought. The only thing he could absolutely confirm was his own existence and summed up his argument in favor of his existence by saying, "I think therefore I am".

We each need to reevaluate all our behavioral choices with the vigor and radical doubt Descartes proposed. Thoroughly question every behavior. Consider the action of jumping in the car to run to the grocery store to buy milk. If we are to affect change in our world, we need to deconstruct and question every part of our behavior. Jumping in the car is an action indicating less than the mindfulness it takes to make well-thought-out, efficient choices. Living our lives more consciously would cause us to recognize the need for milk in a less urgent state. The car's existence may need to be reevaluated. Do we really need to own a car? Does it really need to be the car it is? Could it be a more efficient car? Could we use a more sustainable means of transportation like a bicycle or our feet? Is the grocery store we're going to the most sustainable choice for how to acquire our provisions? Are we going to Walmart or the like, who news media has reported on numerous occasions to use less than fair trade practices and make other than sustainable choices. Are we going to Trader Joe's or the like, who news has reported to sell sustainably sourced products and ensure their employees and supply chain are treated fairly. Is the milk you are choosing to buy from local cows which are treated humanely and is it packaged locally, so there's less fossil fuels involved in getting the milk from the cow to you? Is the packaging the milk will be in, the most sustainable choice? There are estimates up to 80% of plastics never make it to recyclers and end up in landfills. Glass and paper recycling rates are, however, significantly higher. All these questions are a demonstration of the deconstruction of one behavior many people do every day without considering it much at all. It's just the tip of the iceberg of analyzing all our behavior choices.

Sustainable means maintainable, supportable, justifiable, defensible, and viable. So only behaviors which meet these criteria are sustainable. When we consider some of the behaviors so culturally ingrained in our society many of them do not measure up to the sustainability bar. Buying stylish furnishings and ornamentation items to decorate and redecorate a house so large we don't use all of it is not sustainable. These things will end up in landfills. Thoughtlessly supporting, with our buying habits, the chemicals and plastics industries brings them to generate more and more products which won't break down for thousands of years isn't sustainable. Adding more and more agricultural effluent to our waterways in the name of providing for industrialized agriculture and the commercialized packaged foods industry isn't sustainable. So many more behaviors we do, both individually and as a society, don't meet the sustainability criteria. Many people think of their cars as a good thing. However, there will be a point in the future where our children will not be able to have internal combustion, gas powered cars. And that's a good thing. So having an internal combustion, gas powered car is not a sustainable behavior. And we have to find another method to provide for our transportation needs.

The popularity of dystopian literature suggests to me more people than not, feel if we keep going the way we are, things are not going to turn out well for all of us. Complaining about a problem serves no value if one is not also going to offer a solution. There are many people who proposed many different solutions but always consider the source of those solutions. What's the motivation of the entity offering the solution? Consider who is trying to influence our behavior most heavily.

We each have to work hard to counter the efforts of unsustainable commercialism. We all know it's better not to buy into the marketing hype. If I did a man-on-the-street interview and asked people "what are your thoughts on commercialism", I wonder what the response would be. The results might be tainted by where I did my poll. Do I ask people in a mall or other shopping district this question? Do I ask people at a farmer's market? Do I ask people outside the food stamp office? Do I ask people at a national park?

Large companies with considerable financial backing have products to sell and their considerable financial backing is based on selling those products, not saving Mother Earth. Their motivation is to sell products; to convince us we need the products, whether those products are sustainable or not. These companies have a vested interest in convincing us we need to buy things. Even the government supports convincing us to buy things, so we help grow the economy.

On either side of the plastic or no plastic argument we have chemical companies versus nonprofit organizations and individuals. The chemical companies sell products and have the financial backing of a large, profitable, and legislatively connected corporation based on the selling of those products. The nonprofit organizations and individuals have nothing to sell to acquire any financial resources, and, therefore, have little or no financial backing. The chemical companies want you to buy lots and lots and more and more of their products, so they will use all their considerable financial and legislative power to convince you to buy and influence you to use more and more. The nonprofit organizations and individuals have little or no power, financial or governmental, to convince you to not buy anything. Their power to help you accept you don't even need anything and walk away from the hard-sell, sneaky marketing of the financially well-supported corporations.

Sustainability is about each of us taking our own power back from our very economically motivated society and living consciously. Making choices based on our own real needs not the false needs other sources have convinced us of. There is much doom and gloom in the world and we all talk about how the news has nothing good to say. If we seek out the good news, we will find more and more people doing things like volunteering, using less plastic, not fishing in areas already overfished, buying locally raised, organically grown, groceries from local sources keeping those dollars supporting local families and many other good choices. Trust in the goodness of the people around you and that things are getting better. Sustainability is about each and every one of us making every decision wisely to keep ourselves, our families, our communities, and our Mother Earth healthy.

CHAPTER 2 – TRANSPORTATION: GETTING OURSELVES BETWEEN POINT A AND POINT B

Struggles With Cars

Recently, I was presented with the conundrum of "to buy or not to buy" a new (to me) vehicle. Buying a new vehicle is a love-hate thing for many of us, me included, and even more so for a committed sustainabilitarian. Since I work from home, I hate the idea of spending the money on something I need so little. But I love to travel and go exploring, so, since this country has not yet mastered train travel the way Europe and the rest of the world has, I have to have a vehicle to do it in. The tendency our culture has taught each of us to get excited about having new things is also part of the love-hate struggle I have with buying a car.

After many missed opportunities for travel this winter, I had just about convinced myself to break down and do it. But being the sustainabilitarian I am, I really wanted to walk the talk I am talking with all my sustainability speech.

Everyone, all around the world, in all walks of life, see cars, and what they represent differently. Traditionally, people have embraced the car and celebrated the freedom they get from it, but the news and popular culture today seem to indicate people may be rethinking freedom a little bit. People are beginning to accept the reality of the true cost of owning a car and choosing to find alternatives. There is an emerging, informal movement towards car free living. A small but growing number of people are beginning to see more freedom in riding their bicycles or walking to their destinations. People are looking for excuses to get out of their cars.

So, while I was trying to talk myself out of buying a car, my companion's daughter also planned to buy a car. Still trying to convince myself I didn't need to buy a car, he and I arrived at a local dealership, which we know to be reputable and not pushy. We were just going to look at what they had for her, kind of check out what cars they had in stock which might be good for her. On a cold but sunny, early spring day, he and I found ourselves walking around this car dealership peering in windows and reading stickers. We strolled up to a car he thought she might like but then noticed it had a standard transmission. It was right off the list of possible options for her. However, as I considered the vehicle's merits for his daughter, I read the sticker, then walked around it, then peered in the back window. "I puzzled and puzzled to my puzzler was sore." We stood and looked at it a few more minutes as we knew I was contemplating the car issues I described. As I mentioned above it was spring, and spring always reignites everything about my being. It gives me energy and resolve and power. This car was made by a company I know to be as environmentally friendly as internal combustion can be, and of very reputable quality. It had a designation defined by the lowest emissions for any traditional gasoline powered vehicle available. And it was a pretty maroon color. I hate all the non-color cars out there in variations of beige, black or white. Suffice it to say, I found myself hugging the car right there on the dealership's lot. I looked at my honey and said "it may not be right for her, but it seems like it was kinda made for me".

This love-hate thing I have with cars goes quite a ways for me.

My mother has been known to carry on quite a bit about the number of her cars I totaled, or at the very least banged up pretty badly, when I was young. It all started with a great big truck when I had my driver's license for only three months and it got uglier after that. These days I don't bang up cars nearly as much. I like to put that down to experience, but I expect it's also because I willingly recognize my weakness and drive a whole lot less. For many years I've been trying to set up my life to not need a car.

When I was a young woman I lived in London for a time, when I first got there, I loved being able to take public transit anywhere I needed to go. After some time in the city and after quite a bit of exploring, I wanted to venture further afield. I found British Rail could take me to any other city in the country or beyond. It could even take me to some of the small towns, but if I got myself to those small towns, I could do all the sightseeing I wanted at the local train station and village, but then what. I wanted to explore the countryside, so I needed a vehicle. As we all are when we're young, I was impoverished and never ended up getting a vehicle the whole time I lived in London. I never got to explore the countryside, but I did quite a bit of exploring on the train and even got over to France.

Quite a few years later, I was in Sardinia, Italy, however this time I was not living in a major city. I was living in a very rural small town with no train service. Since it was a small town, I was able to walk anywhere I needed to go; my local café, the weekly market in the town center, friend's homes, my job, restaurants, and whatever else there was around town to visit. And again, in the beginning, I was content with the lot I had been given in my happy, little, small-town life. However, after some months in a small town, one can usually say they have fairly thoroughly explored. The traveler in me began to get itchy feet and I knew there was a lot more to see on this island than I could get to on my feet. I found myself getting friendly with anyone I knew who had a car. I got to explore the island quite a bit, but I had to do it on my driver's terms.

All these instances serve to expose me, and I expect plenty of other people's, love-hate relationship with cars. Our whole culture has a love-hate relationship with cars. We have romanticized the idea of the open road and the freedom of getting in the car and going anywhere. Consider how our nation has embraced NASCAR racing as a uniquely American sport.

We have also demonized the car. Manufacturer's advertisements describe how their car makes it more pleasant to be in the car for long periods. Young people of today are moving back into the city centers, so they don't have to deal with car ownership and long commutes. There's now a thing called a walkability score for towns and cities across the nation and the world. It's a number which takes into account many different factors affecting how easy and comfortable it is to live car-free in a given location.

Let's consider some of the financial cost out of your budget for owning a vehicle, things like depreciation, taxes and fees, financing, insurance, maintenance, repairs and the one everybody thinks of first, fuel. What is the cost of your time and energy put into things like maintenance, repairs, parking, and traffic? Now imagine how much it would cost to rent a car or participate in some of the car sharing services, say, one week in the month for some adventuring and exploring.

Consider the environmental cost of what comes out of a vehicle. Even today's all electric cars have an environmental cost. How much petroleum was used to get the raw materials out of the ground and to the manufacturer to get that vehicle from the manufacturer to you? The batteries it uses aren't fully recyclable? Another environmental cost of all of us driving our own cars is the cost to community. There is plenty of documented and anecdotal evidence which shows how much better off people are when they live in a connected community. People on bicycles or walking talk to the people around them. They're not insulated and isolated in a personal private bubble.

I have struck up some of the most interesting conversations with people I've met on trains and buses. Riding on trains and buses gives people time to read, knit, write, learn the language etc.; all the things which take some brain power and we can't do because we have to give brainpower and concentration to driving. At the points in my life when I've been a train commuter, I was able to understand, sympathize with, and appreciate such a broad variety of people.

About a year ago, my companion was rear-ended in our older, but still very reliable and serviceable, second car. Since I now work from home, I jumped at the opportunity to try my hand at car-free living. I talk a big talk about making decisions and changing behaviors based on what Mother Nature needs from us and how we can give her the tools to heal the damage we've done. Living car free was a big step towards walking the walk of all the sustainabilitarian talk.

We live in an area with a really poor walkability score, but it was summer, I have a bicycle and I wanted to try it. I did great with it and had no problem finding ways around car travel when needed.

Until it came to long-distance travel. I found myself faced with the same problem I had been looking at when I lived in London as a young woman. I wanted to be able to get in the car and go roam around the countryside. For example, my children live many states away and I wanted to drive to be able to visit them. My mother lives in Arizona and when I wanted to get out of the cold Northeast and visit her for the winter, I found I was trapped without a car.

Sure, I could fly to those places, but how does one get around when you're there. One has to rent a car or pay a lot of taxi fares. So, I'd broken down and bought a car. However, I made sure it was low emissions (only because I couldn't afford the capital investment of a zero emissions vehicle).

So now I owned a car again, I felt like my experiment in car-free living had failed and I spend time considering why, and the largest reason seemed to be the lack of ability to do any traveling. I rarely use my car when I stay near home, but I'm comforted to know I can make a spur of the moment trip when I want.

So, when I found myself hugging the fabulous car on the dealerships lot, I pondered all of the issues to be considered regarding car ownership. It was a pretty long hug. And then I turned away and said "No, I don't need a car!" But the damage had been done with the sun streaming down on me and spring time infecting my every decision, I went through the rest of the day trying to convince myself I had not already made the decision to buy that car.

Walkability

I walked into a small-town diner and standing at their counter I asked, "Is there a grocery store nearby? Within walking distance?" We had arrived in town on our boat and so anywhere we needed to go, would have to be on foot. I had asked my question, knowing full well the definition of walking distance is drastically different for many different people. I like walking very much, so my definition of walking distance is a mile or two or more. The woman behind the counter at the diner grimaced and with the wave of an arm, she said "Oh no. It's gotta be a mile or mile and a half up the road".

She had waved her hand in the direction I needed to walk so I stepped out her door, pondered the time it would take and how pleasant the walk would be and set off figuring I would find more guidance along the way. I set off walking in the direction she indicated. With no sidewalks and a dusty, uneven roadside, it was slow going. As I walked, I came to a bridge with no road shoulder. This is where I began to question my safety and I turned back. As I returned to my point of origin, I determined the appropriate question was not "is it walking distance?", but more appropriately "is it a walkable trip?"

There are plenty of people who live blocks from a bus stop yet still take their cars. There are also people whose offices are within a mile of a bus stop, yet they still take their cars. What makes one person walk farther to get on public transit and what will cause another person to drive their car even though there's a bus stop nearby? The woman in the diner had told me the trip was not walkable and she was right, but not because of the distance. Walkability is dependent on such a variety of things including safety, weather, pleasantness and ease of the journey (i.e. wide sidewalks or dusty, rutted road shoulders), major intersections and roads to cross, and last but not least, distance.

I walk because it feels good; I want to get some exercise and get some place too. I walk because I'm poor; probably not as poor as some other people, but I choose to spend the money I have on things other than transportation. I walk because I'd rather limit my support and use of fossil fuels; it's a meager contribution, but it's what I can do.

A walkable community also has to have strong public transit because, although the majority of our destinations are near home, there are times when we have to go further afield. For those destinations, I have been a public transit user and lover all my life. I love the idea of getting on a train or bus and making good use of the time spent in transit lost in a good book and or magazine, instead of being stuck paying attention to the road.

In college, I rode the bus to school each day. My home was about a block and a half off the main road. There were nice sidewalks in my neighborhood. I lived in Tucson, AZ, so the weather was always nice. Using the bus was so pleasant and easy, I didn't even own a car. I could get on the bus a block and a half from my home, ride for less than a mile, and then have to get off and change buses. Or I could just walk the mile, then get on one bus and take it all the way to campus. I could walk or take a short bus ride to a grocery store, the library, restaurants, and entertainment. This is the definition of walkability for me.

After college, I lived in London and rode the Tube everywhere. Years later, when I lived in Washington, DC, I rode the Metro everywhere and did my shopping while walking between the Metro station and my destination. I've never lived in New York City but had boundless opportunities to spend time there and love riding on the subway every time. I've happily never used a New York City cab. In Paris, while using the Metro, I was so taken by the beauty of the Metro entrance signs. They are done in an Art Nouveau wrought iron style transporting one to the Paris that is the artistic center of the universe. Strong public transit is part of what makes a community walkable, because it gives citizens a means to travel to those destinations further away from our local neighborhoods.

With all the health problems our society is facing due to lack of exercise, or even an active lifestyle, it seems to me that the government would be more keen to make the infrastructure of our towns and cities more walkable. Based purely on my own anecdotal evidence, obesity rates in big cities (I'm using New York and Washington, DC) seem to be lower than in rural communities, like the agricultural communities of Delaware and Maryland. In big cities, people find cars to be such an inconvenience they're willing to walk and use public transportation. In rural communities, things are far enough apart nobody goes anywhere without their own individual car. In these communities walking is something to be done merely when your car is broken.

There are simply times when people need to use their individual cars or trucks. People who do work with lots of tools or equipment need their cars/trucks to haul those trappings and people who are traveling where public transit doesn't go have to take their cars too. When I was a mother in a minivan, walking would have been a huge challenge for me because of the number of stops I had to make, the challenge of herding kids and the amount of stuff to be carried. I still made every effort to leave the car home. We would walk or ride our bikes to school in the morning. And I'd ride over in the afternoon to pick them up again. I even rode my bike to the grocery store. I had my smallest child in a two-kid sized trailer and groceries went in with him for the trip home. We lived in Arizona and it was sometimes pretty warm, so I think he appreciated sitting next to the frozen foods.

When I worked in Washington DC and traveled to Delaware on the weekends I had to take my car. After calling everyone from bus services to the military base to hotels and anybody else I could think of who might have regular customers going that way I determined there simply is no public transportation between Dover, Delaware and Washington DC. As committed a public transportation user as I am, I couldn't find anything.

My Mom used to live a very short distance from two major shopping centers. Out her back gate there was a bit of a ravine with enough growth to make her yard feel like it didn't open up onto a parking lot. On the other side of the ravine was the parking lot, fairly quiet and it led down to a major intersection. This intersection is where a decision was to be made. You could cross a quiet, 2-lane road to get to a little shopping center with groceries, hardware, clothes, a nice little café and a couple of restaurants, and quite a few specialty shops. Or you could cross a very busy, 6-lane divided highway and then have to cross the 2-lane road to get to a giant shopping center with a bigger grocery store, a department store, multiple restaurants, and endless specialty shops. We all thought of the big shopping center as not a walkable trip, but the little shopping center felt homey, local, and welcoming. They were essentially the same distance from her home, and we frequently walked over to the smaller shopping center yet never made the journey to the giant one. And never found there was anything we needed that we couldn't get at the pleasant, hospitable, little shopping center.

In a good many communities, there are more challenges to walkability than there are challenges to having and using one's own car. We, as a society, will need to tip the balance in the other direction before we can see the changes needed to save ourselves and Mother Earth. Walkable communities have endless benefits, far beyond the tangible benefits of less gas used and more exercise done. When I walk I smile more and stop to chat with and get to know my neighbors, which has been shown to increase security. I take the time to smell the roses and appreciate the world around me. When I walk I feel the happiness and joy that comes from being in the outdoors and experiencing the world al fresco.

What Kind Of Transportation Gives You The Heebie-Jeebies

Our happy, little group sat at an outdoor, street-side table of a downtown restaurant where we met up to begin our vacation together. Our friends had come into Baltimore by plane and we had arrived on our boat and we met up here as it's between the marina and the airport. As they walked up the street coming from the light rail station, I rose from my seat with a big grin on my face, wrapped my arms around them saying "it's so good to see you guys. We are going to have so much fun together".

The beer was cold, the hors d'oeuvres were fresh and the trees gave us full, cool shade. The pleasant place and the good company gave us all back the energy drained by the day's travels in the heat and humidity. We all had a nice time getting reacquainted with friends we hadn't seen in too long.

"How was your journey?" I eagerly asked, knowing she hadn't traveled much, let alone on trains. "What do you think of that light rail train?"

She cringed, describing the heeby-jeeby feeling she got from the train ride she had just taken from the airport. "As we pulled into one station, I saw these kids, and then watched them when they got on our car. They looked so disrespectful and acted so selfish with their loud music and pants sagging too low". Reliving the moment she was on the train, watching those "hoodlums" get on her train car, she shuddered, shook her head and said with determination "not again!", vowing not to ride a train again. She seemed to feel as though her security was in jeopardy.

As she described it, I knew exactly how she felt because I got the same feeling from my taxi ride from the marina. To me, taxi's feel dark, slimy, too closed in, too much beyond my control and absurdly expensive. I had looked up bus schedules and routes, comparing those to the map of where the marina is. I had it all figured out how to get from the marina to the train station where we were meeting our friends. I told my honey, "Look the nearest bus stop is less than a mile away." When it's up to me, I'd walk miles and take a bunch of buses through sketchy neighborhoods before taking a taxi ride. "If we don't feel like having to change buses we can walk just a bit further on to the next bus stop and skip the bus change", I told him. I plan my schedules and activities around the time it takes to walk, so I don't have to take a cab ride. "The whole journey shouldn't take us more than 30 or 40 minutes." Alas, the allocation of how we'd use of our time that day had not been only up to me. My honey and boat captain told me "I really want to take a few more minutes to get this project done. Then we can grab some showers and jump in a cab so we won't be late." The afternoon's work had not allowed us time to make the bus journey I had planned and I could feel the anxiety building in me as I realized I was going to be compelled to take a taxi ride. When I'm with other people, I feel trapped by my companion's unwillingness to walk and forced to do something I don't want to do.

My friend and I got the same heeby-jeeby feeling from two totally different transportation experiences. People won't be willing to use transportation giving them the heebie-jeebies and can have similar barriers to totally different transportation experiences. For my friend, a more individual, less public transit method will have to be provided. For me public mass transit is a very viable solution. I spend time in public places like grocery stores and libraries with strangers. It seems no different to me spending time with them on buses or trains. My security feels more in jeopardy in a taxi than on any bus or train I've ever ridden. Anything we, as a society, come up with as an alternative to individual cars, has to be something people like me and people like my friend are comfortable enough to use. There is no one right answer to our society's transportation ills.

When I lived and worked in Djibouti, Africa, my colleagues all seemed so blasé about jumping in a cab to get where they needed to go. But, after much searching, I could come up with no other option. Since we were coming from a little bit outside town, and there seems to be a communal perception Americans would not want to use buses, the cab companies seem to have pushed out any public transit. So, I had to suck it up and deal with my heebie-jeebies. I didn't like the cab ride so much, and I do love walking so much, sometimes on my one day off a week, in the African desert sun, I would walk the 5 km into town.

As much as I love public transit, over the years there've been occasions to drive a car. If I have multiple stops to make on my journey, I can take the car and do it in an hour, but if I take the bus, it will take most of the day. I may have too much stuff to use public transit, or it's too big to easily haul on public transit. I once needed a 4x8 sheet of plywood and had quite a job wrestling that onto the top of my car. I can't imagine how I would've wrestled that onto a bus. Another barrier to using public transit I've experienced is how it may, simply, not go anywhere near where I need to go. I used to work in Baltimore and Washington, DC and drive home to Delaware on the weekends. I would've loved to take a bus into work, knowing I could use public transit when I got there. But there were simply no options for public transit between my home and my work.

I think about the fact that even I, as a lover of public transit, find reasons to need to take a car. What about people who are more ambiguous about public transit. Or people, like my friend, who are downright opposed to public transit. How is society ever going to convince them to use the bus or train every now and then, let alone give up their cars?

My Mom tells a story of when she had gotten a new job she was very excited about, working at the New York Botanical Garden in the Bronx and lived in New Jersey. From her home to work was about a 20 mile journey, but this was New York City and its surroundings which are known to have great public transit. After some investigation Mom found it would've taken three hours and three or four different vehicles to get there because of where public transit is geared towards getting people into and out of; Manhattan. She would've had to take a bus to get to a train system, then a train into Manhattan, and the subway out to the Bronx, finishing up with a bus to the Garden, taking 2-3 hours to do it all. So instead, every morning and every afternoon, she would spend over an hour, if traffic was good, driving her individual car through some of the most congested streets in America to get herself to work and back. This is an example of how public transit is not able to please all the people all the time and another example of the challenges to getting people out of their individual cars.

My experience living in a London suburb was very similar. A job right up my alley became available in the suburb a few miles around the ring road. Feeling excited about this job, I investigated how to get there on trains, buses, and tube. Being young and poor and living in a city with great public transportation, I didn't have a car. It was an unnecessary luxury considering the expense, and it'd probably be used for nothing more than weekend outings. I poring over the transit schedules and timetables. I determined the only way to get there would be a two hour excursion on three buses through the suburbs or a two-hour excursion on three trains if I wanted to walk a few miles, if the weather was good (this was in London). I ended up not even applying for the job. And I expect mom would not have been so excited about her new job if she was trapped into the two or three hour commute each way. That experience makes me wonder how many other people living in poverty are trapped there by their lack of ability to get to good jobs.

Mom and I's experiences serve to highlight another challenge to getting people out of their individual cars. Cities set up their trains in spokes going out from the center. Getting into an out of the center of cities is easy. But if the city center is not part of your journey, there will be some challenges to overcome. When I had recently moved to Washington DC and had an appointment, and knew this was the first in a series of appointments, just a few miles around the infamous Beltway from where I lived. I could ride the Metro train down into the center of town, to the transfer point and then back out again to where I could walk to my appointment. Or I could drive my car those few miles around the Beltway and be there in less than half the time. I was new living in the DC area and felt nervous and reluctant to drive on the Beltway. For the time savings involved, I bit the bullet and drove to the appointment.

People find, for any number of reasons, our society's current state of public transportation will not serve their needs adequately. People would rather take their own cars than walk to a transit stop, possibly carrying a number of belongings, and then walk on the other end of the ride, again carrying belongings. Urban planning and transportation professionals speak of the challenge of the first mile and last mile in which moving lots of people between popular destinations is easy and cost effective, but getting people to those popular destinations from whatever less frequented destination at which they start or end their journey, is a far greater challenge. Giving people transportation options between a large office building and a location for dining and shopping, or between a stadium and the parking lot a few miles away can be accomplished more effectively through mass transit than individual cars. Parking, both fees and availability, and traffic makes taking one's own car more of a challenge than walking to the public transit.

Ideas for alternatives to each of us driving our own car range from plans reminiscent of Jules Verne, which don't look so far-fetched these days, to Tesla's extravagant and exciting ideas of autonomous buses and PRT (Personal Rapid Transit). However, Even Tesla's autonomous buses are about moving large numbers of people between popular destinations. Getting people out of their individual cars becomes a more of a challenge when you only have a few people to move from one location to another. How do we get people from their houses to their bus stops and train stations and then, on the other end, from bus stops and train station to work and shopping. And how about doing it without having to change vehicles 3 times.

People's barriers to using public transit are broad ranging. My friend and I's similar heebie-jeebies, generated from experiences on opposite ends of the public transportation spectrum, remind me everyone has different needs and motivations. There is no one right answer for our society's transportation problems. There is no one reason a person will walk farther to get on public transit while another person will drive their car when there's a bus stop so nearby. The balance between convenience of personal transportation and convenience of public transportation will have to tip through many different means before people will be willing to change their habits.

CHAPTER 3 -

HOUSING: THE ROOF

OVER OUR HEADS

AND HOMES

A Comfortable, Happy State Of Homelessness

"You want to buy a cargo trailer and make it into an RV?" my honey asked, with head cocked to one side and eyes squinted at me in a quizzical look. "Seems like an awful lot of work. Why not just buy an RV?" he pondered. I explain to him about how I want to be able to sleep anywhere; about how I want to do some stealth camping; about how it will be a cheap place to stay when I visit the kids or do any travel. RVs are not designed to be self-sufficient and off-the-grid.

I tell my honey, "I wanna be a vagabond. There are so many places to see and so little time. I gotta make my life more mobile. I wanna roam and all I need is a warm, dry, safe place to sleep each night." He supports me and knows I'm passionate about my choices, so although he thinks I'm crazy, he helped me anyway.

My ideas along these lines started years ago when my kids were little. I wanted to take them traveling and provide plenty of outdoor living and the convenience of an RV to be able to head out on the spur of the moment. Then, my idea was a campground and lots of outdoor living, but now I want to travel in urban and rural environments too. As time went on, I wondered, "if I have an RV that's self-contained and I don't need to use any services, why do I have to pay for a campground?" If I'm respectful and polite and don't infringe on the rights of those around me, why can't I live wherever and however I want?

A movement of people wanting to live very small and with a lighter impact has begun. These people would rather fill their lives with experiences, not things. One trend in this tiny-living movement is converted cargo trailers. The advantages to this are lighter weight, more easily mobile, and can be used for free, stealth camping. They live in self-contained and inconspicuous vehicles. They stop wherever they want at night, then get up the next morning to go out and find whatever the next adventure might be. I want to live in my converted cargo trailer/RV without the need to seek out a place to sleep each night, or the place I'll wash each day or cook my meals.

A couple years ago, I got the guts to take the leap. I bought an insulated cargo trailer and converted it to a small, off-grid RV. My RV is really small and designed for one person, maybe a couple, if they really like each other, but keeping it very simple and lightweight, there's no reason I couldn't apply the same principles to a larger trailer.

The process took $7000 and about a month. It was only as expensive as that because I bought a brand-new trailer. And due to the high cost of the whole solar set up. I parked my trailer in the driveway and, with the help of my very supportive and capable fella, worked on it out of the garage. I kept it very simple, starting with painting the plywood walls and insulating the underside. Next we added the kitchen, a used countertop and sink on two stock cabinets. I choose to use an RV pump-handle faucet to maintain mindfulness of my water usage. The bed is a mattress on milk crates with lids. This system offers tons of storage space and is reconfigurable to be a couch during the day.

My honey, ever the optimist, says "What about a freshwater tank? Having long since considered this whole operation I shot back "I'm gonna put 5 gallon buckets under the sink. One for fresh water and one for the discharge gray water." The plan was to refill the supply bucket by hand, but refilling a 5 gallon bucket is heavy and indiscreet so, in time I changed over to 4 one-gallon containers which fit into a milk crate. I fill each of my supply containers by hand, but an addition I'd like for the future is a rainwater catchment system and a water filter.

All the cooking and heating is gas powered. My electricity will come from my roof mounted solar panels and be stored in my battery. I'll use my electricity for lighting and charging my computer and my phone, but I'll also use it for incidentals like running a fan when the world gets particularly warm. I happen to like my world to be very warm, so I don't expect all need a fan very often.

My toilet, at the moment is a porta-potty. In time, I'm going to install a composting toilet with a sawdust bucket under a wooden box disguised as a seat.

Many of the people who do this stealth camping thing don't have showers because it's easier to join a nationwide gym chain and just shower there. I happen to be retired military and so I can shower at any base gym in the country. One of the really exciting possibilities for camping with this RV is the possibility of sleeping in the wide-open nothingness of public lands and I have a solar shower bag I can hang on the outside of the RV. I've even thought about putting brackets on the outside of the RV for a U-shaped shower curtain rod, so I can shower outside even when I'm not in the wide-open nothingness.

In trying to keep this non-traditional RV inconspicuous, I have very few windows so it sometimes feels a fairly dark, closed in space. Cargo trailers can have a large ramp door on the back, which, in this case, becomes opening a wall of your home and inviting the outdoors in. I open my ramp door and prop the end on jacks so its level and I have a patio. It's one of my favorite features. And I'm even contemplating a way to make it a screen room.

This whole thing will be an ongoing project because, as I use my RV, my experience with it gives me notions to improve my living situation. There are other people walking similar journeys, by choice or by life's conditions, and we'll continue to share ideas and information. I'll pay attention to their ideas and read the information they make available as were all in the community of people with non-traditional homes. I have lots of big plans for development and I can't wait to continue my journey.

Experimenting In Homelessness

There are lots of people in the world who are, what the government would call, "homeless". These are people who, either by their own choice or by virtue of the circumstances they've been given don't live in a conventional home. These people are living with all levels of income and comfort. Some live in a van traveling and sleeping in a different Walmart or Home Depot parking lot each night, some live in big, snazzy RVs parked in full-service campgrounds, some live in boxes under overpasses, some live in modest, vintage or home-made trailers spending time in many state or federal public lands, some live in boats. Wherever they live, the place they call home doesn't look like what the legal establishment calls a traditional home. By not having a traditional home we are considered homeless. Instead of trying not to be homeless, what if people embraced it. This might be useful for all varieties of people with unstable housing situations, whether they be unstable by choice or by consequence. There's a whole community of people who could be served by off-grid living.

My cozy little off-grid cargo-trailer-converted-to-an-RV is also the beginnings of an experiment in homelessness. Safe, comfortable, affordable housing has become such an insanely expensive proposition that more and more people have to go without. The establishment and all the regulations they've instituted have pushed housing beyond the reach of so many people. If I was looking at an eviction or foreclosure notice, an off-grid RV like this would be a viable option. I want to live in my RV in unity with the homeless or people who can't afford a safe, let alone pleasant home. The most important feature of my RV, to so many actual homeless folks, is that it's a place to feel sheltered and home. A largely overlooked problem of homelessness is the lack of a place to feel secure and safe and home.

There are practical and moral components to my experimenting in homelessness. Why am I doing this? Do you want the "idealist trying to make the world a better place" answer or "cold, ugly, daily living" answer? Both are equally true. I can't afford flights or hotels to do any traveling or get to see my kids who live many states away so this is the only way I can make it happen. And the way housing works today in America isn't working. We all need to figure out we can live with less. I want to show it can be done. My experiment is about showing how sustainable living can be done inexpensively. This means things like no expendables like paper napkins or towels to be bought and then discarded into a landfill and skipping the costly, over-packaged, processed foods in favor of cooking from scratch, or doing laundry in a bucket with a plunger and a salad spinner and a drying rack.

As I present all my high and mighty ideals, I have to accompany them with my homeless experiment disclaimer. I have a job working over the Internet so I'd work throughout my experiment. I usually live pretty minimally and low impact, but because I'm not homeless and living in poverty, my experiment is not an accurate representation. No amount of living like I'm homeless can replicate really not knowing where one's next meal will come from or having a place to call home.

The saying "life is a journey, not a destination" certainly applies to this adventure. This RV and the journey it is, are new to me. I continue to have ideas and read articles with plans for improvements. I have lots of big plans for continued improvement in my head, like the water catchment system I mentioned above, and look forward to the journey.

My off-grid RV and the journey it is, are freedom to me. The fewer things I have, the fewer things I have to worry about losing or finding a way to maintain. I'm not suggesting everyone should live like a homeless person, but this is a demonstration of how we can all live with less. And a statement of solidarity with those who have no choice but to live with less. This is a means of showing the government and commercial establishment "home" is not only a large, expensive to own and maintain house with traditional, on-grid utilities anchored to a piece of land.

What's So Great About Off-Grid Living

I'm sitting on the couch, with the computer in my lap, surfing and daydreaming. My honey walks past me and says "what are you looking at? Tiny houses?" I got caught red-handed when I was supposed to be working. He sat down next to me and I showed him a very ingenious table designed for tiny living. It's a table for four that stacks against the wall, in no more space than a sofa table would take up. How ingenious is that?

With my honey sitting next to me, I continue to click through an array of tiny living articles, with pictures of tiny homes and tips and tricks for living tiny. He's looking at some of the pictures with me and I tell him what I like and don't like about each home we look at. For example, I mention how that one is a nice house, but the electric stove and heating would have to go because it'd take a LOT of solar panels and windmill to run all those electric appliances. The water catchment system on another house is very clever because it filters the water before and after the storage tank. A different one is nice for some other reason, and our discussion goes on like this. He knows I like the idea of living off the grid and tiny.

Off-grid means something different to everyone who chooses it. Some people have installed sources of renewable energy, like solar or wind power generation, and a battery bank in their homes, or simply eliminated their power needs, and are off the power grid. There are people in what is known as the slow food movement, who have giant home gardens and chosen to grow or raise all their own food and barter with their neighbors using whatever abundances they have for anything they can't make themselves. These people are off the grocery store grid. Some have chosen to give up their cars and are off the car grid or chosen minimalism and are therefore off the consumerism grid. Off-grid has come to mean choosing not to participate in some convenience our modern society provides because the price, be it cash or intangibles, is too high.

Choosing to be off the energy grid, for me is mostly about security. Not the sort of security where I'm worried about some thug hurting me or taking my belongings away for me, but the sort of security where I'm worried about poverty, due to any number of circumstances, removing my ability to take care of myself and my family. What if I got to the point where I couldn't afford the house the government and zoning has mandated must be a certain size and must be supplied by on regulated utilities to even qualify as a place where I can legally live. Still needing some place to call home, a small, off-grid RV with some solar panels and a water catchment system solves those problems.

For me, living off-grid means living minimally and being self-sufficient, just you and the essentials. No building codes, no dependency on the power grid, no required monthly payment to the trash collection service or city sewer service. The idea of living off-grid is about simplicity and making conscious choices. Challenging oneself to push the boundaries of the things we can live without, and for each of us, those things are different. My mother and I have such different needs and different means of cooking. I could live without a microwave where she, on the other hand, could live without an oven.

At one point in my life, I had been unemployed for a while, and finances were seriously tight. When the trash collection bill arrived, I had to evaluate whether or not to pay it or to cancel the service. We already had a compost in our backyard and the town provided a vigorous recycling program for those willing to bring it to the collection site themselves, which we already did. Consequently, my "trash" was very small. My family of five filled our kitchen trash can in no less than four days. With my refuse habits on one side of the balance and a trash collection bill on the other side of the balance, I canceled my service. We ramped up our use of recyclables and disposed of what little "trash" we had in the park trash cans, where I took the kids to play. I've since found a good job, my kids have grown beyond taking them to the park to play and I've reinstituted my trash collection service. But the whole experience made me much more conscious of using items whose final destination is a landfill.

The table for four I mentioned earlier, that stacks against the wall in no more space than a sofa table, doesn't meet standardized chair and table height and size requirements, which builders and manufacturers lobbyists have managed to get legislated, would be unacceptable as it doesn't meet those criteria. Our consumerist society has so standardized everything, including home size and supply chain, anything outside those specifications is not commercially viable, or sometimes even legal to be sold. If I want to control the size of my home or how I generate power for it, I should be able to do that suiting my individual needs. Though it is outside standardized power generation methods, penalizing the homeowner who's put up a windmill and a few solar panels seems to me, to be merely the old power generation and transport infrastructure being unwilling to let go of a system that's served them very well financially and move towards a new innovative system that's customizable to suit each of our individual needs.

We live on our boat, and I love living on our boat because it is self-sufficient. Boats are designed with freshwater tanks, wastewater tanks, hot water heaters, refrigeration, and electricity generators including a means of getting electricity from the engines. We have all the things and systems one needs to live off-grid on board. I recognize these are not all needs, but they are wants and they do make life a little easier and more comfortable.

I see freedom to choose off-grid living doesn't mean I have the freedom to infringe on my neighbors rights to peace and quiet by installing a loud, smelly generator. And there are some people who see windmills and solar panels as an eyesore, so that would be infringing on my neighbor's rights to a pleasant view. I should not be allowed to do it to the point where my power generation efforts infringe upon my neighbor's rights to peace and quiet, and I guess that's where legal issues begin to be involved. Who gets to decide when I am infringing on my neighbor's rights? I think the guy who sees windmills as an eyesore is nuts, but we all know beauty is in the eye of the beholder and it's not my place to say whether or not he's nuts.

A long time ago, Kris Kristopherson wrote a song about two people roaming around the country and in the song there's the line "freedom is just another word for nothing left to lose". I remember hearing this as a young woman and being floored by the power in the statement. I'm not suggesting I'd like to bum around the country with nothing but a backpack, but a great many people feel a certain appeal in such an idea. America was founded on people's rights to choose for themselves how they wanted to live. The Pilgrims and all the other settlers found the oppression and conditions in England and Europe intolerable, so they loaded their lives onto boats and went somewhere they could make their own choices about how to live. Today, if a citizen finds the rules and regulations put in place to maintain the old ways and the old infrastructure, to be intolerable, they should be allowed to remove themselves from it. I have a, possibly over romanticized, vision where each household, through wind, or solar, or any other renewable means, generates all the power the individual household needs to operate stored in batteries on site, eliminating the need for any grid at all. Thereby eliminating the need for fossil fuel burning, or even nuclear, power plants.

Human beings are, however, communal creatures and the idea of "community" means people have chosen to live together, sharing resources and burdens. I see our community sharing more burdens and hoarding more resources. Our consumerist society doesn't give people the chance to not spend money on things. Living off-grid doesn't mean living outside a community, like a hermit. There's an idea of a social contract which states members of the community have consented, by the active living in the community, to surrender some of their freedoms and submit to the authority of the rule of the community. Our society's rules govern all aspects of community life, like required amounts of electricity provided to a home, or requiring overly cautious building standards, or compulsory trash collection, begins to cost more than what a person has. When the submittal to the rule of the community has become more powerful than individual's freedoms, to many of those individual freedoms are surrendered.

I'm not suggesting community living is bad or even unacceptable. However, for me, requirements to pay for services I neither need nor want or to live in a home bigger than I want or any number of other overly cautious legislated obligations is undemocratic and dictatorial. Off-grid living is about making conscious choices and taking back the freedom to make the choices each and every one of us finds to best suit our individual situations and needs.

CHAPTER 4 – ENERGY: POWERING OUR LIVES

Taking Energy Availability for Granted

My happy little off-grid RV is all hooked up to our truck and I'm off for another adventure in independence. I designed and built this RV to be able to meet my needs without connecting to the grid and it means I have to be mindful of the resources I have on board, like water and cooking gas and power stored in the battery. I have to have an understanding of how much power I need to run my devices, including basics we take for granted, like electric pumps to move water, and how much my solar panels will collect.

A few days later, it's any average workday morning and like any other morning, my 7 AM alarm goes off reminding me, it's time to go to work. Rolling over I stretch and push my feet flat against the wall of my RV. My curtains are closed so it's still darkish inside. Lying in bed for a few more minutes, my brain starts ruminating on the activities of the day ahead of me. I'm always attentive to how much water and gas I have on hand and I know there's enough of both. I make sure I have enough of both before I go to bed so I don't have to deal with acquiring those resources before I even wash up and prepare myself to face the day. Swinging my feet around, I dig my toes into the fuzzy, warm rug. Standing up, I give one last stretch. I put the water on to boil for washing up and making tea. While the water is warming, I use the cold water for tooth brushing. Teeth cleaned and washed up, I'm dressed and ready to face the day.

Pulling my curtains back, the sun blazes in and I'm glad it's sunny out. Not only does the sun make me smile and fill up my personal batteries, blazing down on my solar panels, it fills up my household batteries. Sitting on my couch, I open my computer to start my workday knowing I'll have enough power to do what I need to do. As I sit, comfortably getting my work done, I feel cozy, content and relaxed knowing all my needs are met.

An hour or two into my workday, I set the computer down and stand up to get the blood moving and make another cup of tea. I look around my intimate dwelling and feel at home and lucky for my happy abode. Standing at my kitchen counter looking out my window onto the asphalt, I'm struck by the disparity in setting from outside to inside and by how amazing it is I could feel so pleasant and content inside my home in... a parking lot. I can live in the wide-open spaces of a National Park, big-box store parking lot, a beach, any truckstop along the way to my next adventure or anywhere I choose.

Sometime ago, way back in the 1990's, I remember talking to my mom, when solar power was becoming more and more feasible and, on a separate research track, electric cars were becoming more and more feasible too. "Mom, I live in Arizona, where we get more sunny days than anywhere else in the country. Wouldn't it be cool if I could cover my whole roof with solar panels and drive a plug-in car? I wouldn't use any fossil fuels at all!" Fast forward to the next iteration of my life and I don't live in such a sunny place anymore. Solar panels have come a long way and so has wind power. I could cover my whole roof with solar panels and put up a wind turbine generator and still not use any fossil fuels at all. With battery technology increasing the storage capacity too... I get so excited about the possibilities, and then so disappointed by the roadblocks. The initial outlay for all those solar panels and the electric car is still quite steep so I'm not there yet... But my off-grid RV is a step in the right direction.

This is not lack of energy availability, it is energy independence and appreciating where energy comes from because it comes from the power generation equipment that goes with each home, whether that home be on land, water or wheels. We've all been spoiled by having all the power we could want readily available, at our fingertips. Never do we have to consider the amount of power we have on hand and ration its use until we can afford to buy more.

Consider the gallon of milk, or the juice or any other staple, in any of our refrigerators. Do we ever run out? Say its first thing in the morning, before you can have your morning coffee or tea, so you have to run to the store and pick up some more milk? If you're anything like me, you make sure the needs for those staples are met so you don't have to go to the grocery store for milk first thing in the morning.

What if electricity came in containers, like our milk, to be purchased at a store and we had to be mindful of how much we have on hand to cover the needs we have for it? The grid was set up because electricity could not be sold in containers and even if it could our democratic society wanted to make sure every person could be delivered the electricity needed to provide for a family's basic needs. However, having "the grid" has made us all take for granted power will always be there. Power always being there was the purpose of setting up the grid. But what if power, or water, or other resources we take for granted, were not unlimited. What if we had to be mindful of electricity usage the same way we have to be mindful of data usage on our phones and other connected devices? Technology is fast approaching a point where every home could have, right in their own yard, the means to provide for all their needs using renewable resources.

Upon considering building an off-grid RV, I needed to know how much solar power I would need. Contacting solar power companies and experts. I asked how much power generation in solar panels I needed to run my 7 x 12 RV/home. The conversation with the salespeople from these companies always began with the question "what devices do you need to operate?" It became clear to me a 7 x 12 RV with a microwave, an electric stove top and is heated with electric needs a lot more power generation then a 7 x 12 RV using electricity only for lighting and charging a phone and computer. After considering the question, this answer seems abundantly obvious, however, even I, who is deeply mindful of these things, had never stopped to consider it. I found myself delving into the whole new world of power usage. I never stopped to consider how much power my refrigerator uses or how much power I need to run my laptop and phone. I had always just plug these devices in and let them run, knowing there would be enough power. So now I had to determine what I needed to run and how long I needed it to run. And that's how I determined how much power generation through my solar panels I would need to provide my RV the required power. Do you know how much power your home takes to run?

There are people up in arms at the idea of the number of jobs that would be lost if power plants and the grid were no longer needed. I submit to you the ice industry of the 19th century and early 20th century. There were a vast number of jobs available in cutting ice up north and transporting the ice to the south. Technology came along and invented refrigeration and all those jobs were lost. This gave those people the opportunity (yes, opportunity... To challenge oneself to achieve something better) to be retrained either in the emerging refrigeration industry or in something they really wanted to be doing, something they had some passion for or would feed their soul or, at the very least, be safer and fill their pockets more. It took a long time and, I'm sure, there was much consternation and angst over the process, however our society evolved and now, no one much remembers the suffering that accompanied that evolution.

What if we retrain all the people who currently work in the fossil fuel industry to work in the renewable resources industry? What if we put all the money we're giving to the fossil fuel industry into figuring out a way to have all the power generation for home, or even commercial building, needs right there on site? People would hate the idea of not having unlimited power but it would make us not take electrical power for granted.

Little Bits Of Energy Saved Add Up To A Lot Of Fossil Fuels Not Burned

The end of the phone charger that plugs into the wall outlet lays, uselessly, on the counter. The other end of the cable is plugged into my honey's device. He plugged it in 10 minutes ago and went about his business. When he came back, it wasn't charging and he wondered why, but quickly discovered the other end wasn't plugged into the wall. As I always do, I had, sometime ago, unplugged the little power sucking box electronic devices use. "Why do you always unplug the chargers?" he says to me with irritation… and then indulgence in his voice but quickly answers his own question. "I know, I know. Because they use power, but it's such a miniscule dribble of power."

All those tiny little miniscules, in everybody's homes and offices, add up to some serious power usage. I don't do it to lower my individual electricity bill. We all need to do it to lower our whole society's electricity demands, and therefore fossil fuel demand. The goal here is to decrease our whole culture's power demands.

It's 80° out on a beautiful, fresh spring day. I get in a car, and the driver has the air conditioning running. Why do you need air conditioning when the outdoor temp and humidity is within a normal, comfortable range? My driver reports "It's easier to turn air-conditioning on then to open all the windows. And having the windows open makes a car so windy and noisy." This person is taking for granted the power used to run air-conditioning.

We've all heard a hundred times "every little bit helps" and "little bits add up". Why don't we heed that? Why does the point seem to go nowhere? What is the psychological science behind people not seeing how individual small efforts can add up to major society-wide shifts in habits?

Maybe people think it's not very fashionable to be conservative with resources. Affluence is portrayed in conspicuous consumption. Our society has developed an odd contradiction between consumerism and conservationism. Even now there are new start-up companies hawking fair trade, ethically manufactured products while at the same time the intended consumers for these companies are the people learning to live with less because they see that it's a more sustainable choice.

As I choose to reuse things in order to save the energy needed to make new ones, make my own cleaners, personal care products and anything else I can think of, and dream of solar panels, wind power machines and living off the grid, there is some derision at my choices. It may be all in fun and I know the people who do it love me, but it's nonetheless, uncool to be a tree hugger.

Cool is not giving a second thought to anything; nonchalant and unruffleable but it's also dispassionate and impassive. When I was growing up, The Fonz" was the ultimate in cool characters. He had no concerns in life. He didn't worry about riding a motorcycle with no helmet or pinching pennies to be able to feed a family. The Fonz would never involve himself with something so mundane and pedestrian as whether or not he should care for Mother Earth.

At the traditional grocery stores in my town, there are piles and piles of over packaged foods. A cool cat like the Fonz only concerns himself with the easiest, most trouble-free (for him) way of getting things done. Even as a tree hugger, I recognize packaged foods and paper products are easier than cooking from scratch and washing dishes. But I also recognize the cost of those things, and not just the monetary cost, to be something significantly higher than I want to pay. We are willing to go to a little bit more trouble to cook out on a grill. If we can't let go of a little bit of "cool" to take care of Mother Nature and pass a healthier world on to the next generation, we won't be looking so cool when our planet is uninhabitable.

We all need to be willing to not just blindly do whatever's easiest. For the sake of making our planet a healthier place, let go of a little cool and unplug the power sucking boxes charging your electronics.

Powering Our Society's Future

Watching a show about whaling and the whaling industry, I was startled by how much used to be done with whale oil. Near the end of the show one of the characters, having heard about how they recently started getting oil out of the ground, was marveling at the amazingness of such a crazy notion. What once was crazy, eventually becomes commonplace, then becomes entrenched doctrine and when it becomes obsolete, people are unwilling to let go of processes and products that are inefficient.

Standing in the garage, someone said "twenty years from now, this will all be obsolete". There are endless estimates about how long petroleum oil will last. I heard a news story recently where a man in North Dakota, where their oil boom has been a big shot in the arm for struggling communities, our society can't just quit oil. It will need to be weaned off oil. And my first wondering was, so let's start weaning. But if we consider where we've come from, we've already started weaning.

There are many people and companies who have a vested interest in not stopping oil production and even ramping it up. We, as a society, need to find a way to change those people and companies over to renewable sources of energy. These people have earned a lot of money and gained a lot of power from oil production. There is much talk about legislating and forcing our society to reduce its dependency on fossil fuels however people are motivated by their own needs. When the people and companies with a vested interest see their needs better fulfilled by renewable sources of energy they will stop producing and using fossil fuels.

Some of our most respected entrepreneurs and most admired thinkers in our society are people who are forward thinking and are planning for where our society will be, way in the future, not where we are now. They are the people who see their needs better fulfilled through other means. People who are making their money in fossil fuels are scrambling to try to find a way to keep their industry viable. Our forward thinking entrepreneurs are working to manage the growth of the industry they know is going to supply power to the future.

I dream of a society in the future where energy will be clean, renewable, and won't cause any damage to our environment. I'm an idea man, and not very good at having the commitment or work ethic to make it happen but there are plenty of people strongly committed to the science of making powering the future of our society with clean renewable energy.

The Big LED Change Over

"You'll never buy another light bulb again" they said.

"You'll create less demand for fossil-fuel-generated power" they said.

"You'll save thousands of dollars" they said.

A few years ago we found ourselves standing in the light bulb aisle at our local big-box hardware store, feeling dumbfounded and paralyzed by the overwhelming range of options. Walking up and down the aisle for half an hour or more, reading packages and comparing prices, color temperature, lumens, watts, hours of life, etc., we found ourselves parked in front of this array of options for simply lighting one's home. Other customers came in, picked out their selections and were gone while we still stood there gazing, incapable of making a decision. I turned and looked at my honey, who doesn't deal well with frustration, getting angry and we decided enough is enough. We walked out of the store empty handed and I'm not sure if it was more frustrating to have spent so much time or to be empty-handed or both. A sometimes forgotten tenant of minimalism and sustainability is to use the things you have instead of loading up landfills with perfectly serviceable products, so we continued using the light bulbs we had for the time being and resolved we'd cross that frustrating bridge when nonfunctional light bulbs necessitated it.

For 100 years, incandescent light bulbs were the only light bulb available, and we all learned the terminology associated with them, and learned to feel comfortable with the color of light they produced. Some years ago, our society started seeing the florescent lights we're all familiar with from our elementary school cafeterias, as well as those oh-so pleasant government buildings, in the form of light bulbs that fit into the lamps we all have in our homes. They were still the same awful color of light as the government buildings, and nobody wanted to use them. Since those first compact fluorescent light bulbs (CFLs) began to supplant incandescent light bulbs, the variety of energy saving options has exploded. The light bulb aisle has become as bad as the cereal aisle for the prodigious array of choices. Many of us are so overwhelmed by the array of options in the cereal aisle, most of them being really unhealthy and overpriced, we simply grab the same box we've always grabbed, which is another opportunity for reevaluating options and making a conscious choice, instead of an unconscious one. Many people seem as overwhelmed as my honey and I by the light bulb options, they end up sticking with what they know, the good old incandescent bulb, even though we know it's not the best option.

I am a big advocate of doing the smart thing, whether that

smart thing be the ingenious ways our grandparents and great grandparents used to get things done or the smart thing being the latest and greatest, technologically advanced methods. About a thousand years ago a guy named William of Ockham, who was an English Franciscan friar, scholastic philosopher and theologian proposed Ockham's razor, saying "Among competing hypotheses, the one with the fewest assumptions should be selected." which for me translates to the simplest solution is probably the best one. I resisted power windows in my car till much later than most people because I didn't like the idea of complicating the simple act of rolling down my window. And I'd still be resisting if I could find a car without power windows. It seemed to me to be just one more thing with the potential to break. I drive a standard transmission car for the same reason; automatic transmissions are just one more thing to break. I'm capable of rolling down my window or changing my own gears but not capable of fixing the automatic versions of these things when they break.

In the old days, houses were designed with double hung windows, the simple and ingenious idea being you could open the top or the bottom depending on whether you wanted to let out hot air or draw in cooler air. People utilized the physics of hot air rises and cold air sinks. They also had transom windows over doors utilizing the same physics. They ingeniously figured out a way to get the job done using the technology available to them.

In the old days we, as a society, determined lighting our houses using the old methods is not the smart thing to do because candles were significantly less effective than we needed. The incandescent light bulb was invented and it served well for many years. Now our society has determined such a means of lighting isn't good because we found ways to create the same amount of light while drawing significantly less power. When compact fluorescent light bulbs first started hitting the market they were pretty universally disliked for the color of the light they produced. I also didn't like the color, but I was an early adopter because it seemed the smart decision both for my pocketbook and for Mother Nature.

It seems to me the smart thing is to buy the bulb with the lowest cost to lifespan ratio. However, let's consider the definition of smart. It means showing intelligence or good judgment. So the next point to ponder is, what is intelligence and good judgment? It seems to be the ability to learn or understand things or to deal with new or difficult situations. So the smart decision for me my not be the same as the smart decision for another person. As a tree hugger, what's smart to me is the financially and ecologically practical choice. Some people may find the aesthetics of the color of the light to be more important. Some people might find the choice requiring the least amount of effort on their part to be the smart thing.

These days we live on our boat and it's older, so it's filled with incandescent lights. We also have the original generator, which is doing well, but we try to be gentle with it and we have such minimal power needs. We also simply don't like to have to turn it on because it's noisy and, although it's only a bit, it still drinks fuel. On a boat, fixtures run on AC and DC so they can run off the batteries so changing over to LED light bulbs also means changing lighting fixtures.

The issue of which light bulb to buy, the one we put off a few years ago when we walked out of the big-box home improvement store empty-handed, has now become more urgent. The less power we can use on the boat, the more we can live without turning on the generator. So we again revisit the question of which light bulb to buy. Putting off the decision seems to have served us since we no longer wonder if CFL or halogen is an option. In the time we were avoiding dealing with it, technology seems to have delivered a clear winner in this battle. Decisively, LED uses less power than all the competitors. So we changed over to LED.

CHAPTER 5 –

WASTE: THE FINAL

DISPOSITION OF

ALL THAT PASSES

THROUGH OUR LIVES

The Great Public Recycling Bin Caper

I've been wondering for years, why every trash can doesn't have a recycle bin right next to it. We all have them in our homes. It's even state mandated. But why not the public trash cans?

On a recent trip to Germany to visit my brother, I was standing in his German kitchen, with its small refrigerator and European design cabinets, looking at three separate bins to dispose of household waste. He explained to me which waste goes in which bin and expressed with great vigor, if the city's waste removal crew found an inappropriate thing in a certain bin they would simply leave the whole bin on the curb and the government also had very stiff fines in place for people who didn't dispose of items properly. Wow, what a great system!

When we first started recycling waste from our homes, some years ago we had to separate glass, plastic, paper, aluminum, etc. I even had a compost in my backyard. My family of five only created a small grocery sack worth of trash to go in a landfill every 2 to 3 days. I still announce this tidbit with pride whenever the subject of recycling or waste comes up. And the town we lived in had a very poor curbside recycling system so we stacked bins in our garage and every few weeks when we drove the hour away to visit grandparents, who had a much more vigorous recycling system in town, we would load all the recycling up in the back of our car and put it in the appropriate bins. Today, my family of only two makes no more waste than we used to.

In years past, when there was less abundance in the world, if a family could afford to throw away questionable milk or a whole apple because half of it was rotted and there was a demonstration they had money to spare. Poor people had to find ways to make the questionable milk useful and cut off the rotted part of the apple, making use of the good half. Waste became a sign of affluence. Waste became something to aspire to. People felt they had really arrived if they had enough affluence to just toss something away and get a new one. Once upon a time, a fat wife was a sign of a rich man. I'm not a fat wife and I don't have a partner, but I could still do better at reducing my waste.

A year or two ago, after a lifetime of listening to his sustainabilitarian mother, my 11-year-old son had a class field trip to the local landfill. If I hadn't converted him before his field trip, he was now totally buying into how bad it is to fill up landfills and bury our trash and why. When he came home, he announced with amazement a very large number of pounds of trash which go into landfills annually and the people running the field trip seem to have done a very good job of helping the kids understand how to reduce that number. He announced he wanted to have a compost bin in our backyard and he was going to join the Green Team club at school. I'm so proud of my environmentally conscious young man.

Just a few years ago, I took my kids to the state fair. We all think looking at the animals, displays of 4-H activities, and the jam, cake, etc. contests are the coolest thing about a state fair because it's the only place you'll find stuff like that. One of the larger, more intricate displays was set up by our government organization for waste disposal. There were trivia games, scavenger hunts, recreations of waterways to demonstrate polluted and unpolluted and lots of interesting information about recycling. Being the sustainability geek I am, I found someone to talk to about some nagging questions I've had trouble finding answers to. For example, is it better to air on the side of recyclables into a trash can or trash into a recycling can? In other words, if I have a piece of waste and I don't know the appropriate bin to dispose of it in, which do they recommend? To my delight, the woman recommended putting it into the recycle bin. She said they would rather have trash go through their single stream recycling sorting system and end up in the landfill, then have a recyclable item go straight to the landfill never having had the chance to go through the recycling sorter. So now all my unknowns are going to a recycle bin. We meandered on through the rest of the exhibit and finished our day at the fair with sunburnt noses and tummies aching from too much fried food. A good time was had by all and on the way home, I chose to spend time reinforcing the lessons we had learned from the solid waste booth. I'm sure they tuned me out, but I'll continue drumming on the sustainability bandwagon and maybe by the time I send them out into the world, they'll get the message.

As we exited the booth presented by the Department of Solid Waste Management, having learned a great deal about why it's important to recycle and reinforce the value of reducing waste, we were presented with a midway filled with booth selling food on paper plates and drinks in plastic cups or Styrofoam. Where is the German mindset to fine people for not recycling in an ecological disaster like this? As my very environmentally conscious young men and I walked along this midway we couldn't help but notice the recyclables on the ground and overflowing from every trash can. Each and every waste disposal point had only one can and it was a can with a plastic bag which would be bundled shot and dumped straight into the landfill. These recyclables would never have a chance to go through the recycling sorter and go on to bigger and better things. Why do public places have trash cans, but no recycling cans?

The Nightman Cometh...Again

The current population of our planet is over 7 billion and we're expected to cross into double-digit billions within this century. As little as 200 years ago there was less than 2billion people and 2 billion peoples worth of biowaste has a very different effect on the planet then 7 billion people, and counting's, worth of waste. There's really only two options for its disposal; dump it and try to feign ignorance of the growing problem, or find another use for it, there by recycling it. Any of the waste disposal methods seem to fall into one of these categories. Mainstream, non-sustainable society today, chooses to dump it and try to pretend it's not there, just like we do with landfills. Plenty of really smart people who recognize the unsustainability of this system and are undissuaded by government regulations kept in place by the persuasive dollars of large-scale industrial waste disposal companies, have put quite a bit of time and energy into determining the value and safe handling of human biowaste.

In our society people have gotten awfully squeamish about ordinary bodily waste and that's made it difficult enough to discuss that we won't be able to change it until we can publicly debate the topic. People don't even like to say the name, toilet. It's indelicate. We call it a restroom. Who rests there? How many people do you know who would comfortably discuss lavatory

construction and biowaste disposal? How are we ever going
to make the kind of drastic, habit altering, architectural and
structural changes needed if we won't even say the word? With
7billion plus people's poop to find a place for, we haven't got much
time.

All the food we eat, at some point came out of the soil and brings
the soils nutrients with it. Humans, as with most animals have
fairly inefficient digestive systems and a great many of our
nutrients go right out the other end. But our modern system
of waste disposal ensures those nutrients don't go back into
the soil. Instead, they are contained and sanitized to the point
where any nutrients that might have fed some remote soil are
so contaminated with chemicals, we wouldn't want to put it
anywhere, let alone be used as a means of feeding any land.

I've had a little bit of opportunity to travel to some destinations
tourists don't often frequent and I was in Eastern Africa on an
expedition to bring some medical help to remote villages. The
most formal toileting facility we came across was an outhouse
with a cement floor with a hole in it. All business was done in
the squatting position. I never quite got any more information
about where the hole went but there wasn't any smell. There were
plenty of spiders and other creepy crawlies who probably just
appreciated the place to get out of the blazing sun. I sometimes

wonder if the Africans preferred not to use them because they preferred the tribal ways of their community, like walking out behind a bush and digging a hole. There were precious few bushes and we Americans were pleased to have the cement-floored outhouse.

My experience in the African desert, where we never once wasted good clean drinking water on any toileting habits, caused me to reconsider the way we "civilized people" dispose of our unsanitary waste. Who was it who decided it would be a good idea to contaminate safe, clean water for our waste disposal? The person who decided it was probably rich, as taking resources for granted is a luxury of those who have plenty. There are lots of people on this planet who'd be thrilled to have drinking water as clean as the stuff we use to dispose of our waste.

Many thousands of years ago, when people were still hunter gatherers, they probably pooped in the woods and just left it, as many animals do. As we began to live in communities, the communities would have recognized this to be an unacceptable solution and people might have figured out if they bury it, it didn't smell. After about a year of decomposition human waste becomes a safe fertilizer so the buried feces would decay and the people in the community probably even recognized crops grew better in areas used for waste disposal. However, after a relatively short period of time, the area around the community would be well saturated with many small waste disposal sites. Each citizen wouldn't be able to find a spot in the woods to be able to do their business where someone else had not already and had to venture further and further afield. The next logical progression might be outhouses. In areas where populations were too dense to allow for enough outhouses or places to dig outhouses, the waste still had to be removed from the area and dumped somewhere.

Before waste was so neatly whisked away in our clean freshwater, individual households had to find another means of removal. People either had to carry it away themselves or could pay someone to remove the matter deposited in the chamber pots and dispose of it. This task was usually done overnight owing to the task's less than pleasant ambience, hence the gentlemen who performed this excrement collection service were called the nightman. Throughout history, and across cultures, there has been employment available in removal of human waste and in some cultures the nightman would even buy the excrement from households as it could be used as valuable fertilizer. Regardless of who paid who, the service would collect the chamber pot contents and take it somewhere, often covering it with soil leading to the term night soil.

There has, undoubtedly, always been people who are too lazy to dispose of the night soil themselves and too poor to pay the night man to dispose of it. In cities or towns, people might simply dump their chamber pots in a gutter with running water reasoning that it would be washed away with the other street trash. People probably recognized streets were cleaner after a good soaking rain and the excrement in this running water became the closed running water pipes where we dispose of our excrement today.,

Human waste, just like cows, pigs, horses, etc, is nitrogen rich, nutrient dense and completely safe when handled properly. People have been very creative with finding a means of disposing of human waste, most often using it as fertilizer. And even in the age of sanitation, with our understanding of diseases like cholera, many people are turning to composting toilets and have recognized human waste decomposes and becomes a safe fertilizer after a relatively short period of time.

Back in the day, when the nightman would come to collect the nightsoil or dung or feces or excrement or stool or muck or droppings or kak or waste matter, he'd be able to sell it as fertilizer. He get more money for rich people's poop because, in theory, they had a better diet. It may have even gotten mixed in with all the other animals' biowaste and spread on the fields altogether.

Today people are trying to come up with all manner of means to recycle our excrement, to see it as a resource, instead of a problem to be disposed of, the way the nightman, or dung gathering provisions, did once upon a time.

One of my favorite leisure activities on a weekend morning is to sit in the sun with a good book and a cup of tea. Recently, I was partaking of this leisure activity, when my honey asked me what I was reading. The book I was reading was "The Humanure Handbook" and I paused to consider what his reaction might be before I spoke. He knows me well enough to not be too surprised but I wondered if I felt like opening that can of worms with him. He's a pretty forward thinking guy, but not quite as unconventional as myself. After my pause, I reported the title of the book. He's pretty worldly and takes my unconventional ideas in stride and even thinks some of them sound pretty clever. His reaction was to pause himself, furrowing his brow and squinting his eyes like he was searching his memory banks. "About the composting toilet? he said rhetorically "sounds alright, like a good system? A less complicated version of the septic tank". As I thought about his assessment, people have been composting their waste underneath their rural backyards for most of this century, since the dawn of flush toilets.

Returning to the definition of sustainable, not doing anything we can't continue to do forever, we need to wonder where all 7-8 billion, and more, of us will put our waste. It may be an indelicate subject; however it needs to be addressed. Water is a precious and limited resource and if we all keep pooping in it, it'll all be pretty well contaminated. Nutrients are also a precious resource we can use for any number of applications that some very smart people are figuring out. We won't be able to continue using freshwater to dispose of excrement and blatantly disregarding the value of the nutrients in it forever, therefore it's not sustainable. So let's put our squeamishness aside and address this issue.

Let Them Eat a Plant-Based Diet

Those infamous words "Let's just grab a Burger" were uttered as we were out doing weekend chores. Everyone has been there; things are taking way longer than expected, people are getting cranky because fuel tanks are running on empty. We were away from home, had too many more chores to do and were hungry so we broke down and resorted to the quick and easy solution.

I found myself sitting in a fast-food burger joint, thinking to myself "This doesn't even taste good to me" The burger was pretty typical and all the people around me seemed to be liking theirs. After a few moments of pondering my meal, I thought "if I didn't like carrots or walnuts, I wouldn't eat them. Why am I eating this?" And it's really bad for me, to boot.

I was eating it out of expediency. Convenience foods are, oftentimes, not very good for us and pretty expensive for the amount of nutrition they contain. The American commercialized food industry has put a lot of money and effort into convincing us we like high added sugar, high added sodium, low nutrient and cheap-to-produce foods and we bought it, hook, line, and sinker. Fast food burgers are well placed on that list. Nutrient dense foods are what people now seem to be calling superfoods and none of them seem to be meat products. Why was that burger more convenient then a salad or a veggie soup. Because it was readily available.

The day before I'd been to a meeting and was reminded how we all need to be true to ourselves. We all have to do what serves our own needs best, so I made a decision then and there to be vegetarian. I determined I wanted to eat healthier, for both me and Mother Nature.

My husband, as with many people, liked burgers and was a committed meat eater. I wasn't making a decision for the family, just me, although this meant I wasn't likely to be cooking meat for the family anymore. If I didn't like carrots or walnuts, I wouldn't cook with them. I didn't want to cook a meal I couldn't or wouldn't eat, no matter how much the other people I'm cooking for want it. There's too many other options. And many of them are way healthier

We're all well aware there is famine and starvation, both in the US and around the world. When the news reports of organizations trying to address these issues, they don't talk about delivering meat products, they talk about delivering rice and beans. All those food aid trucks being delivered to natural disaster sites and places around the world experiencing famine are carrying bags of beans and grains.

When I was a young, new mother, I received nutritional aid from the government through a program called WIC (Women Infants and Children). They gave me coupons each week for specific items deemed important to ensure myself and my children were getting adequate nutrition. There were no coupons for any meat products but there was plenty of plant-based proteins like beans and peanut butter. The government wanted to make sure we all got our protein but they also needed to do it cheaply. I learned from this plant-based protein is cheaper than non-meat-based protein.

As a young person growing up in the west, I remember driving across the wide-open spaces Westerners are so proud of, past feedlots filled with cows, shoulder to shoulder, standing in the muck, and smelling the painfully pungent odor accompanying those feedlots. I would think about how awful it must be for them. People would tell me, "They're only cows, they don't mind". I just couldn't ever bring myself to accept such a thing. I remember thinking, "Don't they get sick of standing next to the same neighbor all the time? What if they don't get along with the neighbor cow?" "I don't care who or what you are, that smell is awful. And those poor animals have to live with it 24/7. How does anybody think that's okay?"

When I moved to Delaware, a very agricultural state, I started driving past farms with rows and rows of long, low buildings with big fans on the ends. When I learned they were chicken houses, I flashed straight back to those feedlots full of cows and envisioned them with roofs. And people told me "Chickens are really stupid animals. They don't mind". I didn't buy such a line any more than I did for the cows.

After a good many years of being a vegetarian, I started contemplating the cows in the feedlot and the chickens in the long, low chicken houses and how is not the animal's natural state. These animals are domesticated and so they don't have a wild state, but all living things have a state of homeostasis where they are unstressed by internal or external conditions. When I ponder the cycle of life and that everything has to die and that death will feed another life, I found myself wondering why it was more immoral to eat the cows and the chickens than it was to eat the broccoli and the wheat. Farmers mow down acres and acres of living wheat plants all the time. Now, we could get into a deep philosophical argument about nervous systems and the nature of pain, but the point is a living thing died so we could eat.

So why is it unhealthy? It is not, it's just unhealthy after we industrialized it. It's just not sustainable. It won't take meat to feed the masses.

Our planet gives us abundant food resources and there is tremendous amount of the nutrition we need to sustain life and thrive in the plants grown through the cooperation of Mother Nature. Processing that nourishing food through a cow before we consume it, means the cow will get and use the energy in that food. Because the cow has used the food energy, it will not be available for the person who eats the cow. Some people are worried about making sure there's enough food for all the people on the planet, a planet with more and more people every day, planet fast approaching double digit billions in population. If ensuring there's enough calories and nutrients to keep all those people healthy and fed, we ought to be mindful of the calories and nutrients ticket used by the cows.

Human animals are omnivores and a little bit of healthfully, sustainably raised meat in the human diet, doesn't seem unsustainable to me, however the burgers the American commercial food industry has done such a good job of convincing us it's the best thing available for expedient meals on the go are so overly abundant in our diet that we've come to a point where we're putting calories into cows while people around the world are starving and people right here at home are malnourished.

CHAPTER 6 -

MINIMALISM: A STYLE

CHARACTERIZED

BY ABUNDANCE

OF EXPERIENCE,

SIMPLICITY AND JOY

Undoing sentimental attachment

With another successful day on the water behind us, filled

with swimming and family and fun, and with the waning sun reminding us all things must come to an end, we sadly pulled into the marina. With the boat tied up and tucked back into her slip, our day's guests collected all their towels and clothes and coolers and children, and we all lingered on the dock for our long goodbyes and the beautiful sunset.

With everybody gone it was time to clean up the carnage of the day. I stepped inside and looked around, baffled so few people could make such a big mess in such a tiny space. Greeting me was hors d'oeuvres all over the table, cans and cups filled with all manner of beverages, and wet towels in piles everywhere.

The first thing I needed was a place to hang the wet things, so with an arm load of sodden towels and no place to put them I stepped out onto the deck, pondering where to hang them. She's a little boat with lots of beautiful wood trim, totally the wrong place to hang wet towels.

I said to my honey "There's no place inside to hang all these to dry. If we hang them on the rail out here, do you suppose they'll stay put? It's kinda breezy out." He gives me a look of "you worry too much" and drapes them over the rail, "where would they go?" I guess he's right and I also guess I don't have a lot of other options.

With the question of the disposition of the wet towels answered, it's back inside for me and more cleanup. I so love having everybody over and the time on the water but…wow, the mess generated is enough to look like a herd of elephants passed through. So, while he cleaned outside, I cleaned up inside and made dinner. By the time we were fed, and after a day of sun and fun, we were ready for some couch time followed closely by some bedtime.

Too tired to deal with the towels on the rail, and, really, …where would they go… we went to bed.

But the next morning, we figured out where they would go.

All the towels we had hung over the rail last night were now gone. They had blown over the side in the night and the bay had taken them away.

Do you have any things you purchased on a special vacation, or given to you by someone dear, or maybe it reminds you of a special occasion when you felt loved and happy? Do you have any other household items, which to anyone else would appear to be trivial and unimportant?

One of those towels that went in the drink was one of those items which would appear to be trivial and unimportant to anyone else but to me, it was a treasured item. I'd had it for many years, and it reminded me of a good adventure in my life. Granted, it was just a beach towel, but I was attached to it and Mother Nature saw fit to remove it from my life. Immediately upon the discovery of its loss, I felt upset with myself for not holding it closer to me, so I could prevent its loss. But I then promptly reminded myself holding it closer, would mean not using it and storing it safely in a closet, never to be seen or appreciated.

When I think about sentimental items sitting in boxes in storage units or in attics across America and the more stable cultures in the world, I wonder about their value in creating joy for the people who keep them? Does the joy of having it trump the cost, both emotional and financial, of keeping it?

Until well into my adulthood my Gram lived in the same house she had lived in when her children were very young. It was a big, beautiful, Victorian style home as filled with love as it was with the trappings of its packrat owner. And 50 years in her house had every corner and cabinet filled with the paraphernalia she always said she "might need some day". I loved seeing my kids play in the same places I remembered playing as a kid. When her children grew and moved away her house was always the place where everybody came back to for family gatherings.

It was at those cherished family gatherings around Gram's big family table, where so much family folklore is shared and passed on, the conversation, undoubtedly, at some point, would turn to Gram's packrat ways. We were all around the table for an impromptu Sunday brunch gathering. One of my cousins had brought donuts and coffee. My Uncle and Aunt, who lived up the street, brought an Entenmann's coffeecake they had in the fridge. Someone had gone out to get bagels and Gram always had cream cheese in her fridge. With plenty of breakfast goodies spread across the table and family of all varieties dropped by to enjoying the bounty of goodies and family time.

Gram sent at the table, her mobility impaired in her waning years, but still so in her element and brimming with the joy of having her big, beautiful family around her. Somebody asked if she had any plastic silverware and she described exactly where in the basement it was. My feisty aunt, rolling her eyes announces with mockery and love, "She probably put it there 10 years ago." We all giggle as Gram good-naturedly takes the ribbing.

"I knew they'd be useful someday" she defensively fires back.

She was a good sport about the teasing we all gave her but she never changed. I remember, even as a young person, thinking, "but when it's time comes to be useful, you won't know where it is or be able to get to it because it'll be buried under all the other "useful items". But sometime she managed to pull off a zinger.

Her attic was like stepping into the gateway to Narnia or The Secret Garden. There were dusty old lamps, half eaten, tweed woolen suits hanging, a few sparse pieces of furniture and endless boxes each holding marvelous unknown treasures and amazing adventures.

When Gram got older and it was time to move her out of her big house, the task of emptying her house fell to her children, who had grown up there and knew their mother's ways very well.

When it was time to clear out the attic, they found unopened boxes, right where they were put, in that very spot, when they moved in, fifty-something years ago. In all the years Gram lived in her house, she never needed whatever was in those boxes. So why did she box up and move whatever was in it all those years ago?

I wonder about people, like my Gram, whose reaction to poverty, whether it be emotional or financial, any form of living without, is to hold tight to a great number of the things that come into their lives. For me, poverty or living without makes me not want to have any stuff to trap me into space and resources to maintain and store that stuff. What do people like that do when they lose a treasured item, like that towel was for me?

Being the vagabond I am, I've reinvented my life numerous times throughout the years. These have been times when, through my choice or the circumstances given to me, I've gotten rid of everything and started over. As I look back at those occasions, I can't think of one item I wish I hadn't gotten rid of. Those items were obviously not valuable to my long-term joy. They were important at the time, but in the grand scheme of my life I've remembered and treasured the experiences I've had much more than the things which passed through, most of which are long forgotten.

We can undo sentimental attachment in our digital life by letting go of clearing out unused digital clutter. Many of us are using the same email accounts we've been using since the dawn of email. Each of us created a Yahoo or Hotmail account and have been using it ever since. And how many of us have emails stored there from the dawn of email. Do you really need your Amazon receipts from 2003 or your craigslist listings from 2007? Use the search feature in your email account to find and delete all your old craigslist contacts, online shopping receipts and whatever else you don't need from all those years ago.

Undoing sentimental attachment in our homes is what is most commonly addressed when contemplating minimalism.

Decluttering our homes can be the most difficult, but also the most rewarding. Having traveled a bit in my life, prior to my embrace of minimalism, I have souvenirs and stuff collected from around the world. These items are similar to the towel I lost. I'm attached to the memories they hold and the look of my home with these internationally interesting pieces in it. I will use the lesson I've learned from the loss of my towel to remember these items are just things. Loved ones collected around the world are more important than things.

There is just as much clutter in many of our offices as in our homes. In our jobs, we often keep files and emails dating back years, back to when we started our jobs or even to when we started in our field of study. We've looked at the filing cabinets in our offices and told ourselves we need to sort and dispose of all that stuff. But then we go home at the end of the day and don't have to look at it anymore, so it continues to wait patiently to be sorted and disposed of. Ask yourself "when was the last time you needed any of the files from that filing cabinet?" Devise a plan to get rid of the largest majority of the items without sorting. Suppose you unceremoniously toss anything with a date, over one-year-old. Or maybe you could toss anything relating to a previous customer, who is not a current customer. Find a way to remove this anchor around your neck, preventing you from the joy of minimalism.

We can even undo sentimental attachment in our thoughts. Being a writer of poetry, I love the idea of poetry readings. I have sentimental attachments to remembering the poetry I've written so I can recite it at the drop of a hat. I don't keep phone numbers in my head, most of us don't anymore, because I can readily find them when I need them, so why should I keep in my head the poetry written down and carefully stored. I can't actually see myself going to a poetry reading or standing up in front of a crowd to recite my work, but I do like the idea of it. I will be realistic about my poetry and its eventual disposition.

Allow me to help you collect experiences. It's a bit like counting your blessings. The experience of my children, unconventional parent as I am, shows me the beauty of adding something wonderful to the world. My experience of reinventing my life numerous times through the years show me how self-sufficient I am and how I can handle whatever the world throws at me. The experience of my military career showed me how other people around the world live, at both ends of the income spectrum and how lucky I am, as a woman, to be born a US citizen.

The experience of losing my treasured towel has only served to increase my desire to not have any attachment to any things.

I think about my possessions; my stuff. At this point, I've been living in the same place for some years and I'm feeling my inevitable itchy feet. I have things to part with. How many of these things can I live without? Can I live better without them? The minimalist mindset is to collect experiences not things. Joy is in all the experiences collected and the people we've shared those with.

Quarterly wardrobe changeover

The weather has finally changed and the past week or two have felt downright spring-ish. With sunshine blazing and temperatures finally reaching into the range where this committed, winter hater begins to feel like there is some warmth in the air, I tentatively step out onto the deck without a jacket and declare "there are definite signs of spring out here." The beginnings of spring reignites every fiber of my being. I feel the same as the quiet, patient seeds, waiting in the soil all winter. As they begin to warm, they poke their little heads through the soil and change their clothes from their brown seed coverings to a fresh, spring green.

This weather change finds me pondering a clothes change also. Every Spring I go through the clothes in my closet, pulling out my wool sweaters and affirming "I don't care if it gets cold again. I'm not going to wear turtlenecks anymore." This year's weather change has been no different, so my quarterly wardrobe changeover is underway.

Being the minimalist I am, I keep a very small wardrobe in my closet, so after three months of wearing the same 6-8 outfits, I'm ready for a fresh look. I could go buy a whole new wardrobe, but I can't afford it (none of us real people can) and there's no guarantee the clothes would still look good after a washing and no assurance I'd feel comfortable in them throughout an entire days wear. If I pack my wardrobe away every three months, I don't have to go out and buy, to get a fresh new look. I just take it out of the box it's been waiting in and I know the clothes fit, are comfortable and look good on me. I know I like them and I get a whole new wardrobe without spending the time and money to buy it. I haven't been able to totally talk myself out of the excitement of getting new things, I've just chosen to do it in my attic.

When I make the move that sets this spring's wardrobe adventure in motion, my honey is sitting at the kitchen table lost in his paperwork. Pulling the attic door down and opening its ladder is always accompanied by a vaguely irritating, high-pitched sound of the metal springs holding the door up, being stretched. Hearing such a telltale sound, my honey picks his head up and turns to look at me. "What ya doin'?" He knows he will soon be enlisted to help on the receiving end of whatever I'm getting down. Excitedly, I tell him it's time to get out my spring clothes and pack away the winter ones. He rises from his paperwork and meanders toward the attic door as I climb the ladder. Again, in keeping with my minimalist ideals, we don't have tons of junk in the attic, so it's easy to find the ugly, black, plastic trunk I store the off-season wardrobe and shoes in. Sliding it across the floor delivers more unpleasant noises. As I lower the trunk through the gaping hole in our hallway ceiling, also known as the attic door, my honey knows me well enough to be waiting at the bottom of the ladder to receive the trunk.

Once the trunk is down to the floor and into the bedroom, I take everything out and spread it across the bed. I have every pair of shoes and piece of clothing I own, no matter what season it is appropriate to, spread out around the bedroom. As I look at the winter clothes in the closet, I think about which one I like and how often I wore each one of them in the past few months. As I look at the spring clothes on the bed, I try to remember which ones are comfortable and which ones will get worn during the season to come. This is the part of my ritual where deliberating on minimalism and sustainability fills my thoughts. Part of my ritual, and I have to admit, the introspective part, is to analyze what gets to stay in my minimalist wardrobe. You see, a minimalist wardrobe is very, very small, even smaller than my already very small wardrobe. I'd love to be able to let go of enough clothes to have my week-long vacation wardrobe, be my entire wardrobe. We all have washing machines. Why do I need more than a weeks' worth of clothes? I don't have any special occasions I have to have snazzy outfits for. The most dressed up we ever get is our Friday date night, and even that is pretty darn casual. I go to work each day in ordinary pants and shirts, not even all the way to business casual. So, I don't need any professional looking ensembles.

So here I stand, surrounded by all the apparel I own, and I have to decide which ones I will continue to own and which ones go into the charity box. If I had a nickel for every piece of apparel I've put in a charity box throughout my life… Well, let's just say I'd have a whole lot of nickels. But on the other side of such an equation, some of the best quality and most favorite pieces of clothing I own came from secondhand shops. One of the most freeing statements I ever heard said "if you don't absolutely love it, why keep it." Why am I keeping sweaters stretched out at the arms or waist? No amount of wishing will make them unstretched again. Why am I keeping pants that are too small for me? When I get skinny enough to fit into them, they will be out of style or

I'll want to reward myself with a new pair. Why am I keeping business suits? I don't need them for my current work and I never want to do the sort of work where I will need them again. I have to admit, I have one "interview outfit" hanging at the back of my closet. I never wear it but I'll keep around as long as I have the space. The hardest things to get rid of are the things with some memory attached to them. I have a really thick, wool, gray, hand-knit sweater I bought in Greece from a funny little man who looked very old and made me smile when he told me I was beautiful and proposed to me. The memory of it still makes me smile, but I haven't worn the sweater in years. It's just too woolly and misshapen, as many hand-knit sweaters are, but I seem to be unwilling to part with it.

Why do I go to the trouble of whittling down my wardrobe? What is so sustainable about not having so much clothing, etc. After World War II, Americans were taught buying things was good for the economy. We learned an excitement and thrill about having new things. Many people in our society have learned these lessons too well and now spend money and time they don't have feeding an addiction to shopping. Even as a young person, I remember wondering why there has to be economic growth, and why economic stasis is a bad thing. All that growth has to end somewhere because we live in a finite world. We can't continue to expect growth, forever. We, as a culture, need to unlearn "the thrill of shopping." We need to replace it with "the thrill of uncluttered lives."

Each time I look in my closet and I see clothes I don't wear regularly, a little piece of me feels stressed out because I'm a clutter-phob. The extra effort put into finding the things I'm looking for, wondering where to put all these things I don't love and wondering how I'm going to pay the bills generated from buying all these things I don't love causes me anxiety. The more things I don't use taking up space in my life, the less space there is for things I do love. As less and less space in my life is taken up by things I don't love, I feel more and more joy because the tension is gone. As I am more joyful, I am more the person I want to be and happier in my skin.

So, when all is said and done the 2 ½ feet of hanging closet space and four drawers I have is loosely filled with only the clothes I'll wear for the next 2 to 3 months, until it's time to take out my shorts and open toed shoes. I feel such a sense of peace as I look at and think about the orderly, organized, capsule wardrobe I've created. But now I have to get my honey to help me put the big ugly black plastic trunk full of winter sweaters and boots back up in the attic.

Putting My Money Where My Mouth Is

I was given the opportunity to put my money where my mouth is when we moved from our smallish house and onto our 36 ft boat. I talk a big talk of living a minimalist life and not having things to get in the way of the essentials but now I have to actually remove all the last few things from my life, permanently. I have to find a way to get rid of all the things in the house that won't be going to the boat. As I think about the things in the house, they remind me of the good adventures I've had in the past, places I've traveled, time spent with my children and loved ones, meals cooked for my family. Yes, I'm even having trouble letting go of stuff as trivial as some of my kitchen utensils.

I've always been very diligent about getting rid of things I don't use but I actually use these things. So these things have memories and are useful. As I sit here writing this, I'm wearing a bathrobe I've had for over 20 years. It's in good condition and I wear it a couple times a week. But I can't see myself wearing it on the boat; in my new life, which I'm very glad I have the opportunity to live.

When I visit my mom, I, and like to think my siblings too, appreciate seeing things we remember from our childhoods. I find myself wanting to give that to my children. Then I tell myself, the world is a different place. I want to be the very sentimental mom, of the old days, who saved every art project, homemade gift, book and toy from the kids' childhoods, but I want to be a vagabond even more. Hopefully, my kids won't be expecting me to be the kind of grandmother who hauls out treasured items filled with memories for my grandkids. The minimalist in me doesn't even want to be that kind of person. But I feel sad to let that go.

A few years ago, I came to a realization that throughout my entire life, I had never lived in any location for more than six years. So, I've already let a good portion of my past and my children's treasures go. I have now lived in the house I'm in for six years. And I happen to be moving again. I've already let go of a lot of the goodies from their childhood and I have to find a way to let go of the rest. I even want to set the example of minimalism for my children. I have a gazillion pictures of all the treasured moments from their past, something our grandmothers who stored everything didn't have. I will have my pictures to be able to show my grandchildren and tell them the stories of when their parents were young.

When I was younger some of my moves were on buses or airplanes and there was no money to be able to ship belongings so I took with me only what I could carry. It's been a good many years since I had a seriously cleansing move like that but the budding minimalist in me envies what that young person was able to do.

Some of us might use the excuse "you were younger than" but moving like that isn't about being young.

Just a few years ago I had an excited phone call from my mature but very healthy mom, "I accepted the job in the Galapagos!" she reported with zeal.

My immediate answer was, "Yay for you!" And then my mind began to reel through all the questions I'm sure her mind had already reeled through. The first one to answer was "Are you going to sell your house?" That would determine a lot of the answers to all the rest of the questions.

She was 67 and had lived in her home for around 20 years. She had already done a good job of depositing all her children's belongings in her children's homes, since we had homes of our own, so she didn't have that burden, but she certainly had plenty of her own, shall we say, stuff. She's been a recovering packrat for many years, recognizing the pointlessness of being a packrat but paralyzed by the uncertainty of what to do with all the stuff, she was faced with the decision to sell, donate, or toss everything or store everything before she went. Storing everything before she went meant the cost of storage and worrying about her house either sitting empty or being rented for a year. She boldly decided to sell everything. I was so proud of her. She ended up not selling every last thing. Some of her more treasured items came to myself and my sister with the understanding she would want the back someday, and her brother stored her car. The up shot of the story here is, at the age of 67, she whittled her life down to nothing more than she could carry in a backpack and went trekking off around the world. She ended up spending over two years in a comfortable state of homelessness. After she got back from the Galapagos, she spent about a year driving around the country and visiting with friends and family and we all welcomed her into our homes for extended periods and were pleased to have her. As she eventually decided she wanted to have a place of her own, she settled into a more traditional home but very quickly realized this was not right for her. She sold it and headed out for more adventures.

From her adventures I learned, and I hope she learned too, we don't need to have things, just a place to put our head down at night and most of us want a sense of home, but it doesn't have to be big or even built on a piece of land. With my big move to the boat, I'm hoping I will never again live anywhere any bigger, so storing a bunch of my belongings is even silly. If I never again own a traditional home I'll be just fine with that, but I'll have to learn to hold on to the memories without having the things to remind me of them.

CHAPTER 7 -

CONCLUSION - LIVING

CONSCIOUSLY

<u>Dreams of a Big Windfall</u>

All of us have dreams about winning the lottery... Or a rich, unknown, relative passing away leaving us a fortune... Or any number of ways one might come into a large sum of money. On a long unpleasant car journey I recently had to make, I let my thoughts drift. This journey is one I have to make often. It's long, tedious, boring and the radio reception is questionable at best. On my way home, I was dreaming up ways I could make this regular trip less unpleasant and when a mind is allowed to wander, like it does through hours on the freeway, one thought leads, progressively, to the next. This journey is so objectionable because of the loneliness and isolation I feel is so overwhelming. There's no sense of hominess. Home is where the heart is so why can't it be in a little RV in an unknown town, as long as we feel love. My thoughts about making this trip more pleasant were of the small, comfortable RV I'd like to have so I could, at least, have a little of my own comforts and a place to feel at ease, when I make this awful trip, and the thoughts of my RV, spontaneously, led to winning the lottery.

Oftentimes, when people dream about what they would do if they won the lottery, thoughts drift to world travel, luxurious cars and big homes. I have the same dreams as anyone else but being the sustainabilitarian I am, my thoughts are of all the same things, in smaller more, sustainable forms. My lottery dream is of a small, homey, RV with enough solar panels and a windmill to stay off the grid indefinitely and a vehicle to pull it. NOW that would be my ideal sustainable life. I'd live in all the cities where I have family and friends I'd like to spend more time with. And move on when the weather no longer suited my fancy.

As I look around my world and my life, I don't think it's very sustainable. But I also have to remember everything is relative. I'm quite confident I do more than a lot of people. But I also know I could do more. All I have to do is just a little more than what I did

ELIZABETH KELCH

yesterday.

A few weeks ago, a friend, who I had lost track of for many years, and with whom I recently reconnected, was sitting at my kitchen table with me, We were deeply lost in discussion about all the things we had experienced throughout the years since we last talked. My kitchen table is in a sunny corner of my house, and the blazing sunshine was streaming in, bringing hints of spring with it. We had so much to tell each other and were eagerly sharing many details of how each of us had grown and changed in our maturity. At a pause in the conversation, she wondered aloud which trashcan to put her used teabag in. As I pointed to the correct bin, she mused how funny it was all of us have more than one trashcan in our kitchens these days. I wondered why all trashcans in the world don't have a second bin right next to them. And the conversation turned to how much the world had changed since we knew each other in our youths. Even the people who aren't on board with human caused climate change have made a few, minor changes they wouldn't have dreamed of years ago.

I proudly got to show off my homemade lotion and she is impressed by my commitment to the sustainability cause. I tell her all about the changes I've made to make my life more sustainable and how passionate I am about it. She said "your life looks pretty normal." When my friend told me my life looks pretty normal part of me is glad my life looks normal because if I'm too far out in left field people will not see me as a role model. Part of me is hurt my life looks pretty normal because I want to look more like I'm making the changes we all need to be making. So people can see these changes are not unreasonable and even quite feasible.

There are some behaviors I just can't bring myself to change to the more sustainable methods. I have, what my children call, "nose issues", which means I sniffle and sneeze a lot. When I was younger it was much worse and I've just learned to tolerate it because I like the idea of taking drugs every day of my life even less. However, this means I use a lot of tissues. I am unwilling to give up my boxes of disposable tissues. I gave up paper napkins many years ago and even paper towels are gone for my repertoire, but disposable tissues probably aren't going anywhere anytime soon.

I tell my friend all about my dream of a small, homey RV and how it's my lottery winnings dream. I paint the picture of what my very sustainable life in my very sustainable RV would look like for my friend. But I also tell her, since I haven't hit the lottery yet, the life I have now is all I can do and it's good enough for today. If everyone made even as few changes as I've made, think of how much better off the Planet would be. So what I've done is good

ELIZABETH KELCH

enough for today.

Progress, Not Perfection

My friend turned to me, empty can in hand, his face probing. I recognized he was asking where to put it. Pointing, I reply "the recycling bin is in there." We live on a cozy, little boat and it would be silly for him to squeeze past me to deposit it appropriately, so I take it from his hand and put it in its proper place.

He screwed his face up in a question and looked at me again, "You keep a recycling bin on board?"

Knowing my friend sees eye to eye with me on environmental concerns. I say "Of course I do. Don't you?" Does he really not separate his recyclables from his trash on his boat?

He replies "There's no place to put it in the Marina.", as if I didn't know that.

I tell him, "We bag ours up and take it to a recycling bin."

Everybody recycles at home and believe safeguarding the environment is a really valuable cause. We wouldn't think of not recycling at home, so why wouldn't those same sustainable ideals and habits apply when people are away from home? Society is making progress, but we're not all living in balance with our environment… yet.

Since we live on our boat full time, we see the Marina on Monday mornings after all the weekend boaters have gone home and this Monday morning, as every other Monday morning in the Marina, I walk past the overflowing dumpster in the parking lot filled with glass, plastic, cardboard and plenty of other recyclables. Our Marina, like many public places, has only one refuse disposal bin because the businesses who maintain those bins only want to pay rent and emptying fees for one container instead of two. And as I walk past, every time I walk past, I'm deeply saddened by the number of recyclables I see in it, going to the landfill.

There is quite a bit of news these days about a little company called Tesla, so much so it's become part of our popular culture. Just a few years ago people wouldn't have taken any notice of a company doing the kind of work they are. Does anyone remember or have ever even heard of a company called SunPower? They were doing work in Solar Power back in the 1980's and nobody took notice because sustainability was not on people's radar. The popularity of the electric car is another testimony to the progress our society is making. Toyota's Prius first came out in 1997, and it's so popular it's become one of the mainstays of their product line. People are getting the idea things have to change, we can't continue to do things the way they've always been done.

But change is hard and comes slowly. Maintaining the status quo is much easier. It takes a tremendous amount of energy and effort to make the changes we all recognize are necessary in our habits. People are recognizing the importance of changing the way we do things and making those changes happen, slowly but surely. There are hard-core, even militant, environmental activists out there who want to force those changes on our society immediately. But meaningful change people can buy into, and won't push back against, takes time. There's a lot of very smart people who have put a tremendous amount of energy and thought into what motivates people to change. Psychological research is being done all the time regarding what it takes to make people change.

When I was younger, I used to lay in the sun to try to get more tan. A great many of us used to do the same thing. We all called it a healthy glow. But as time passed more of us than not have gotten the message to wear sunscreen and hats. This societal change is still in progress, but has already been in progress for 20 or 30 or more years. Now, I wear a wide-brimmed hat to shade my face, neck and shoulders. I sit under umbrellas and, taking my cues from cultures that of grown-up in the world's desert climates, I cover up with a light airy long-sleeve shirt. It's taken this long for our society to come from that point to where we are now and we've still got a ways to go.

As a young mother, I hated the idea of planning meals because I wanted to be able to be spontaneous. I wanted to have a bunch of groceries in the house and just pick out whatever I was in the mood for on a given day. But too often I'd get busy and have not decided what the meal was going to be until 5 or 6 o'clock in the evening. At that point it's too late to make a meal requiring time to boil dry beans or defrosting ground beef so we ended up eating out far too often. As time went on I began to be aware of which groceries I needed to buy to make what meals. If at any point in the next two or three weeks I'd want to make a chili I should make sure there's a bag of dried beans in the closet. That revelation was the beginning of my willingness to plan meals. The next evolution came when my family started doing some traveling and camping. I had to decide what we were going to eat and take along the groceries for that meal plan and nothing else. Camping is not the sort of event where you'd want to drag along lots of stuff you're not likely to need. My meal planning skills have now evolved to where I decide what I'm going to want to make for the next week, buy the groceries for those meals, and I can be spontaneous within the bounds of my meal plan. I'm still not able to plan what meal happens on what night, but the practical side of me has decided I can plan my spontaneity a week out and then over the next week I can pick any one of those meals I have the groceries for.

My change in sunscreen wearing habits and meal planning practices didn't happen overnight. Over the past 20 or 30 years my choices have evolved, making one small change at a time. Are we asking ourselves, as a society, to do this thing, we as individual people, are well aware is very hard. Would we ask ourselves to change all the infrastructure of our homes, all our eating habits and transportation methods overnight? We all recognize it's much too big a change to happen quickly and without a catalyst. What if we could agree on one issue to focus all our energy on, like the moon shot efforts of the 1960s?

Those people succeeded because a bunch of people were focused on one goal. Currently, our trouble is we all recognize the problem, but we don't agree on the fix for it. The goal is a little bit more esoteric than the moonshot was. We all agree we want to clean up the environment, but there's a lot of factors involved in environmental clean-up and people seem to disagree about the best way to get to our goal. Many people want to find solutions to industrialized food problems. Many people want to develop alternative energies. Many people want to cultivate options for zero emissions transportation. All are viable means of fixing our sustainability problems, but if each of the people who wants to do each of those things is focused only on the issue they see, our societal strength is divided and spread too thin to make the powerful impact we want to see.

What was the catalyst for the small changes mainstream people are making? What made people start recycling? And start carrying reusable grocery bags to the store? And start driving lower emissions vehicles? What makes climate change deniers not change? The cynic in me believes it's motivated by financial gain. They're saying "I've made a bunch of money from living unsustainably and I'd like to continue making money so I'll be very rich and powerful. The future be damned." Years ago, there was a small but vocal minority of granola-eating, hairy-legged, Birkenstock-wearing tree huggers speaking up and making sustainable choices. As time passed "those crazy hippy's" ideas became more and more mainstream and our numbers have grown by leaps and bounds. Even some of the most die-hard conservatives I know are putting a recycling bin on the curb next to the trash bin. Or making other sustainable choices they would not have even considered 20 or 30 years ago. I'd like to think people are seeing we need to make some changes in the way we treat our Mother Earth.

I have to remember, years ago every trash can was filled with recyclables and they ended up in a landfill, just like the dumpster in my Marina. Today, more and more of our refuse is not getting mindlessly dumped into the landfill; more and more people are consciously choosing to use less electricity and water, farmers markets selling locally grown organic produce and animal products are popping up everywhere. Not just the out-there hippies are choosing to conserve. I know every one of us is doing what each of us can manage at the moment, and as time passes each and every one of us will be able to manage more and more sustainable habits.

Responsible choices include stewardship of the environment. When you see trash littering the ground, do you stop and pick it up? More people are picking trash up today than did in years past. When you clean your house are you choosing to use less toxic products? The proliferation of non-toxic cleaning products suggests more people are mindful of the harmful chemicals going into our environment. I recently heard soda sales are down drastically from their peak. This suggests to me people are recognizing soda is not a sustainable product, or even a smart choice, and making wiser decisions. There are many little hints and signs not making the news, but indicating we are moving in the right direction. With our continued efforts to make our world a cleaner and healthier place, we can all appreciate the fruits of our labor. And know we left our little corner of the world better than we found it for our children and grandchildren.

Next Steps

Thank you for reading my eBook. Please leave me a review. I'd love to know you thoughts about my work and the ideas in the book.

If you liked this book, you may also like the works posted on my website elizabethkelch.com

www.ingramcontent.com/pod-product-compliance
Lightning Source LLC
Chambersburg PA
CBHW030656220526
45463CB00005B/1807